Advanced Redox Flow Technology

Scrivener Publishing
100 Cummings Center, Suite 541J
Beverly, MA 01915-6106

Publishers at Scrivener
Martin Scrivener (martin@scrivenerpublishing.com)
Phillip Carmical (pcarmical@scrivenerpublishing.com)

Advanced Redox Flow Technology

Edited by
Inamuddin
and
Tariq Altalhi

Scrivener Publishing

WILEY

This edition first published 2024 by John Wiley & Sons, Inc., 111 River Street, Hoboken, NJ 07030, USA and Scrivener Publishing LLC, 100 Cummings Center, Suite 541J, Beverly, MA 01915, USA

For more information about Scrivener publications please visit www.scrivenerpublishing.com.

Wiley Global Headquarters
111 River Street, Hoboken, NJ 07030, USA

For details of our global editorial offices, customer services, and more information about Wiley products visit us at www.wiley.com.

Library of Congress Cataloging-in-Publication Data

ISBN 978-1-119-90479-3

Front cover images supplied by Pixabay.com
Cover design by Russell Richardson

Set in size of 11pt and Minion Pro by Manila Typesetting Company, Makati, Philippines

Printed in the USA

10 9 8 7 6 5 4 3 2 1

Contents

Preface **xi**

1 Membranes for Redox Flow Batteries **1**
Hridoy Jyoti Bora, Nasrin Sultana, Nabajit Barman, Bandita Kalita, Neelotpal Sen Sarma and Anamika Kalita
1.1 Introduction 1
1.2 Membranes Used in Aqueous Organic Redox Flow Batteries 6
1.2.1 Classification of Membranes Used in Aqueous Organic RFBs 6
1.2.1.1 Nafion-Based Membranes 6
1.2.1.2 Microporous Membranes 7
1.2.1.3 Anion-Exchange Membranes (AEMs) 7
1.2.1.4 Cation Exchange Membranes (CEMs) 8
1.3 Membranes Used in Non-Aqueous Redox Flow Batteries (NARFBs) 10
1.3.1 Stability of Membrane in Diverse Solvents 10
1.3.2 Ionic Permeability and Selectivity 11
1.3.3 Ionic Conductivity 12
1.3.4 Swelling 12
1.3.5 Mechanical and Chemical Stability 13
1.3.6 Cycling Performance 13
1.3.7 Classification of Membranes Used in NARFBs 14
1.3.7.1 Dense Membranes 14
1.3.7.2 Dense Ceramic Membranes 17
1.3.7.3 Porous Membranes 17
1.4 Ion-Exchange Membranes or Ion-Conducting Membranes for RFBs 18
1.4.1 Cation Ion Exchange Membrane (CEMs) 23
1.4.2 Anion Exchange Membrane (AEMs) 23
1.4.2.1 Preparation by Condensation Reaction of Ionic Monomeric Compounds 24

1.4.2.2 Preparation by Polymerization of Vinyl Monomers 24
1.4.2.3 Preparation from Conventional Polymers 24
1.4.2.4 Preparation by Plasma Polymerization 25
1.5 Polymer Electrolyte Membranes 26
1.5.1 Membrane Properties 27
1.5.1.1 Ion Exchange Capacity 27
1.5.1.2 Chemical Stability 27
1.5.1.3 Thermal Stability 27
1.5.1.4 Mechanical Property 27
1.5.1.5 Ionic Conductivity 27
1.5.1.6 Vanadium Ion Permeability 27
1.5.1.7 Water or Electrolyte Uptake 28
1.5.2 Transport Mechanisms 28
1.5.2.1 Proton Transport 28
1.5.2.2 Vanadium Ion Transport 28
1.5.2.3 Water (H_2O) Transport 29
1.5.3 Membrane Preparation 29
1.5.3.1 Cation-Exchange Membrane (CEM) 29
1.5.4 Anion-Exchange Membrane 30
1.5.4.1 Polysulfone (PSF) 30
1.5.4.2 Poly(aryl-ether-ketone) (PAEK) 31
1.5.5 Amphoteric Membranes 31
1.5.6 Porous Membrane 31
1.5.7 Polybenzimidazole (PBI) 31
1.5.8 Polyacrylonitrile (PAN) 31
1.6 Amphoteric Ion-Exchange Membranes 32
1.7 Protonated Polybenzimidazole (PBI) Membrane 34
1.8 Summary 38
References 39

2 Electrolytes Materials for Redox Flow Batteries 51
Mandira Mitra, Sandip Kundu, Mousumi Layek, Subhodip Mondal, Ujjwal Mandal and Bidyut Saha
2.1 Introduction 52
2.2 Overview of Redox Flow Battery 53
2.3 Measurement of the Capacity of the Redox Flow Battery 54
2.4 Formation of Redox-Active Constituents for RFB 56
2.4.1 Inorganic Redox Flow Battery 57
2.4.1.1 All Vanadium RFBs 57
2.4.1.2 Zinc/Bromine RFBs 58

2.4.1.3 Tin/Bromine Redox Flow Battery 59
2.4.1.4 Iron-Chromium RFB 62
2.4.1.5 Polysulfide-Bromine RFB 63
2.4.1.6 Titanium-Manganese Redox Flow Battery 64
2.4.2 Organic Redox Flow Battery 65
2.4.2.1 Quinone-Based Redox Active Materials 67
2.4.2.2 Tempo-Based Redox-Active Materials 68
2.4.2.3 Redox Active Materials Based on Alkoxybenzene 70
2.5 Hybrid Electrolytes Used in a Lithium Redox Flow Battery 71
2.6 Levelised Cost of the Redox Active Materials 73
2.7 Conclusion 74
References 75

3 Zinc Hybrid Redox Flow Batteries 83
Srijita Basumallick
3.1 Introduction 83
3.2 Zn Electrode and Dendrite Formation 85
3.3 The Electrolyte 86
3.4 Effect of Temperature 87
3.5 The Membrane 88
3.6 Hydrogen Evolution Reaction 89
3.7 Conclusion 89
References 89

4 Zinc-Bromine Hybrid Redox Flow Batteries 91
M. Ramesh, J. Maniraj, D. Santhosh Kumar and S. L. Pradeep Kumar
4.1 Introduction 91
4.2 Electro-Chemical Energy Storage 92
4.3 Redox Flow Batteries 93
4.4 Zinc/Bromine Flow Batteries 95
4.5 Types of Zinc-Based Hybrid Flow Batteries 97
4.5.1 Zinc-Sulphur (Zn–S) Hybrid Battery 97
4.5.2 Zinc-Nickel (Zn/Ni) Batteries 97
4.5.3 Zinc-Sodium Hybrid Ion Batteries (ZSHIBs) 97
4.5.4 Zn-Ion Batteries (ZIBs) 98
4.6 Electrochemistry of Zinc/Bromine Deposition 98
4.6.1 Electrochemical Performance 98
4.6.2 Reduction of Dendrite Deposition 99
4.6.3 Bio-Mass Electrocatalyst 99

4.6.4 Surface Chemistry 101
4.6.5 Effect of Zinc Utilization 101
4.7 Applications of Zinc-Bromine Hybrid Flow Batteries 102
4.8 Future Challenges 102
4.8.1 Electric Vehicles 102
4.8.2 Energy Management 103
4.8.3 Size and Cost 103
4.8.4 Safety Measures 103
4.9 Conclusion 104
References 104

5 Zinc-Cerium Hybrid Redox Flow Batteries 107
Raghvendra Mishra and Rajendra K. Singh
5.1 Introduction 107
5.1.1 Redox Flow Batteries (RFBs) 108
5.1.2 The Basic Concept of Redox Flow Batteries 108
5.1.3 Progress and Challenges in the Redox Flow Batteries 110
5.1.4 Types of Redox Flow Batteries 111
5.1.4.1 Aqueous Redox Flow Batteries 111
5.1.4.2 Nonaqueous Redox Flow Batteries 114
5.1.4.3 Hybrid Redox Flow Batteries 115
5.2 Zinc-Cerium Hybrid Redox Flow Battery 117
5.2.1 Working Principle of Zn-Ce Redox Flow Cell 117
5.2.1.1 Components of Zn-Ce Redox Flow Battery 118
5.2.2 Factors Affecting the Performance of Zn-Ce Redox Flow Battery 120
5.2.2.1 Temperature 120
5.2.2.2 Electrolyte Flow Rate 121
5.2.2.3 Current Density 121
5.2.2.4 Charge Conditions and Cycle Life 122
5.3 Summary 122
Acknowledgment 122
References 123

6 Vanadium Redox Flow Batteries (VRFB) 127
Abrar Hussain, Muhammad Tahir Khan, Samad Yaseen, Ata-ur-Rehman and Syed Mustansar Abbas
6.1 Introduction and Overview 128
6.1.1 Working Principle of VRFB 130
6.1.2 Main Components of the VRFB System 132
6.1.2.1 Electrodes 132

6.1.2.2 Electrolytes 137
6.1.2.3 Membranes 140
6.1.2.4 Bipolar Plates 145
6.2 VRFB System as Compared to Other Energy Storage Systems 146
6.3 Recent Research on VRFB 148
6.3.1 Positive and Negative Electrodes 148
6.3.2 Electrolytes 149
6.4 Conclusion and Perspective 149
References 150

7 Vanadium-Based Redox Flow Batteries 161
Anurag Tiwari, Shishir Kumar Singh, Dipika Meghnani, Raghvendra Mishra and Rajendra Kumar Singh
7.1 Introduction 161
7.2 Redox Flow Batteries (RFBs) 162
7.2.1 The General Structure of RFBs 163
7.2.2 Working of Redox Flow Batteries 164
7.3 Types of Redox Flow Batteries 164
7.3.1 Iron/Chromium 165
7.3.2 All-Vanadium 166
7.3.3 Vanadium/Bromine 167
7.3.4 Bromine/Polysulfide 167
7.4 Vanadium Redox Flow Battery (VRFB) 168
7.4.1 Working Principle of Vanadium Redox Flow Battery 169
7.4.2 Role of Different Components in VRFBs 170
7.4.2.1 Role of Membrane 170
7.4.2.2 Role of Electrolyte 170
7.4.2.3 Role of Electrodc 171
7.5 Applications of Vanadium Redox Flow Batteries (VRFBs) 171
7.6 Summary 173
References 173

8 System for the Redox Flow Technology 177
Shrabani Barman and Ujjwal Mandal
8.1 Introduction 177
8.2 General Construction of Redox Flow Battery 179
8.3 Energy Capacity 180
8.4 Optimization 181
8.5 Classification of RFB Based on Active Electrolyte 181
8.5.1 Inorganic Redox Flow Battery 182

8.5.1.1 Vanadium Redox Flow Battery 182
8.5.1.2 The Iron Redox Flow Battery (IRFB) 184
8.5.1.3 Polysulphide-Bromine Redox Flow Battery (PBBs) 186
8.5.1.4 Zinc-Polyiodide Redox Flow Battery 187
8.6 Organic Redox Flow Battery 190
8.7 Membrane-Less RFB 193
8.8 Semi-Solid RFB 193
8.9 Conclusion 194
References 194

9 An Overview of Large-Scale Energy Storage Systems 199
Silas Saka
9.1 Introduction 199
9.2 Progression of Energy Storage Method 201
9.3 Categorization of Energy Storage System 204
9.3.1 Mechanical Energy Storage 204
9.3.2 Thermal Energy Storage 213
9.3.3 Electrostatic and Magnetic Energy Storage System 215
9.3.4 The Electrochemical Energy Storage System 219
9.3.5 The Chemical Energy Storage System 228
9.4 Implementations of Energy Storage Systems 236
9.5 Commercial Prototype of Energy Storage Systems 236
9.6 Environmental Repercussions of Energy Storage Systems 236
9.7 Energy Storage Guidelines 237
9.8 Blockades and Effective Solutions 237
9.9 Future Prospects 237
9.10 Conclusion 238
References 238

Index 245

Preface

Advanced redox flow technology has sparked interest in bulk energy storage due to its flexibility in design, safety in operation, efficient energy storage, and near-zero environmental impact. The technology has been extensively developed and tested at a range of levels in recent years, demonstrating its applicability and usage. As energy becomes a greater global concern, redox flow technology must be considered as a possibility. Issues such as energy shortages and rising air pollution pave the way for renewable energies like solar and wind energy, which have been extensively analyzed and evaluated in recent years. These renewable sources, on the other hand, are intermittent and frequently unpredictable regarding availability, resulting in low-quality output energy and a negative influence on grid stability. To date, diverse types of energy storage systems have been designed for various purposes, each with its own set of benefits and drawbacks. In recent years, redox flow technology, particularly vanadium redox flow, has progressed substantially. Experiments at various scales have been successfully carried out, proving the viability of redox flow technology in bulk energy storage applications. In addition, innovative redox flow technologies that offer more energy storage per unit mass or volume of the storage device and are cost-effective have attracted a lot of interest.

This book presents redox flow technologies completely, from their basic principles to their applications. It provides a thorough review of the craftsmanship in the subject, as well as the potential for future advances.

Chapter 1 reviews the implementation of various membranes in redox flow batteries by highlighting different research around the globe along with the membrane properties applied in different batteries and their impact on overall performance.

Chapter 2 delineates the various aspects of the modern fields of redox flow batteries. It summarizes the different types of flow batteries (like Vanadium RFB, Fe-Cr RFB, Zn RFB, etc.), and the various opportunities for using flow batteries in different ways.

Chapter 3 deals with some important features of a zinc-bromine flow battery with an introduction to promising features of a flow battery. It elaborately discusses the problem of dendrite formation on zinc surface and its prevention by the use of surfactant and using nano Zn composite electrodes.

Chapter 4 focuses on reducing the dendrite growth by proper methods that will make the zinc-bromine hybrid redox flow batteries attractive with their low costs, long lifetime, and high energy efficiency. This chapter also suggests research has to be steered to extend its applications for better batteries for mobile devices.

Chapter 5 discusses the design, operation, and performance of a zinc-cerium flow battery. Also, the general introduction and types of redox flow batteries are discussed. The details of the components as well as factors affecting the performance of redox flow batteries and their advantage over other flow systems are also discussed.

Chapter 6 covers the working principle, the main components of the VRFB system, and a comparison of VRFB with other electrochemical energy storage technologies. The recent developments of VRFB are briefly explained and the most crucial learning research fields are discussed, as well as suggestions for future development.

Chapter 7 includes the concept of vanadium redox flow batteries (VRFBs) as a kind of energy storage device with a cost-effective scaling strategy discussed in detail. The VRFBs are implemented as high capacity, long cycle life, and negligible self-discharge energy storage systems. The working, advantages, future aspects, and limitations of VRFBs are presented.

Chapter 8 explains the different redox flow technologies that are being used at present. Working protocols of redox flow technologies and their advantages and disadvantages are explained. Most interestingly, this chapter gives an overview of how redox flow technologies will become an alternative energy storage option in the near future.

Chapter 9 discusses various types of energy storage systems, their advantages, disadvantages, limitations, and applications. Additionally, the impact of these energy storage systems on the environment, energy storage guidelines, barriers and solutions, and prospects are discussed.

1

Membranes for Redox Flow Batteries

Hridoy Jyoti Bora[1,2], Nasrin Sultana[1,2], Nabajit Barman[1,2], Bandita Kalita[1], Neelotpal Sen Sarma[1,2] and Anamika Kalita[1,2]*

[1]*Physical Sciences Division, Institute of Advanced Study in Science and Technology, Paschim Boragaon, Guwahati, Assam, India*
[2]*Academy of Scientific and Innovative Research (AcSIR), Ghaziabad, India*

Abstract
Because of ever-increasing energy demand, the necessity of storing energy on a huge scale becomes significant in order to coalesce renewable energy sources into the electricity grid. Conventional batteries store energy by exploiting metal-based reactions, which are limited due to their lower efficiency. Redox Flow Batteries emerge as the foremost in the case of energy storage; however, the high cost of these batteries hampers the widespread implementation of this technology. Membranes utilized in redox flow batteries accomplish several significant tasks as they govern the performance along with the economic viability of the batteries. Membranes are also utilized as the separator to avoid cross-mixing of electrolytes, yet allow a route for ions to transfer in contemplation of completing the circuit. Various properties enroot membranes, such as ion conductivity, stability, and low water consumption along with ion exchange capacity. Due to this, the fabrication of inexpensive, chemically and mechanically stable membranes considering the redox flow batteries magnetized the researchers. This chapter discusses the implementation of various membranes in redox flow batteries by highlighting different research around the globe and also focuses on the polymer properties applied in different batteries and their impact on overall performance.

***Keywords*:** Membranes, redox flow batteries, electrolytes, storage, porous, voltage efficiency, energy, grotthuss mechanism

1.1 Introduction

With the ever-changing evolution in the advancement of energy-related technologies, living creatures are the ones that benefit most. Every

**Corresponding author*: anamik.kalita01@gmail.com

Inamuddin and Tariq Altalhi (eds.) Advanced Redox Flow Technology, (1–50) © 2024 Scrivener Publishing LLC

development encompassing energy is followed by the speedy growth of societal forces, along with enhancement in the quality of human life. Mankind has encountered different revolutions in energy. First was the utilization of fire such as firewood for cooking; after that came the use of fossil fuels, viz. oil, coal, etc., followed by the exploitation of coal. In the past century, the use of oil on a large scale bought about the foremost revolution in industries. The technology of Redox Flow Battery (RFB) was first developed in 1970, by the U.S. National Aeronautics and Space Administration (NASA), and they utilized a redox couple, Fe^{+3}/Fe^{+2} and Cr^{+3}/Cr^{+2}, respectively, over the cationic and anionic periphery [1]. The proposed prototype was hampered by the limits of active materials and deteriorated capacity, and there was therefore a high need for further progress of mixed electrolytes to eradicate the problem. Even in the present day, this invention in the area of research and development has shown exponential growth. However, the evolution from hydrocarbon fuels to renewable energy resources is increasingly widely encouraging the advancement of storage for vast-scale electrical energy to modify the recurrent energy produced through renewable resources into a stable and transmittal power. Accordingly, RFBs, due to their appositeness for the storage of vast-scale energy, turn out to be the hot topic in research for the expansion of advanced systems for the storage of energy utilized in various types of applications. In 1949, an idea arose to overwhelm the cross-contamination by employing the same elements with variable states of oxidation on both sides of RFBs [2]. This was realized by the University of New South Wales, Australia, in 1980, for fabricating VRFB, i.e., the vanadium redox flow battery that is often considered the best suitable system so far, where vanadium in its variable states of oxidation is present, having higher oxidation states on the cationic sides, such as V(IV) and V(V), and lower oxidation states on the anionic sides, such as V(II) and V(III) [3]. Subsequently, the VRFBs become the focal point in RFB research and advancement technology, intending to enhance the efficiencies by developing membranes, electrolytes, electrodes, etc. Numerous VRFB-related exertions will be enlarged in well-along sections. The main constituent of RFB is the IEM, i.e., ion exchange membrane, that separates cationic and anionic electrolytes and with the help of which, during charge/discharge cycles, couples undergo electrochemical oxidation and reduction reactions, and the IEM facilitates the passage of charge carriers to sustains electroneutrality, as shown in Figure 1.1 [4, 5]. The main feature of RFBs is that they can separate both energy and power. The stack present in the batteries can control the power and energy stored in the separated reactants. Therefore, one can increase the efficacy of these batteries by optimizing all the conditions to make them relatively easy and

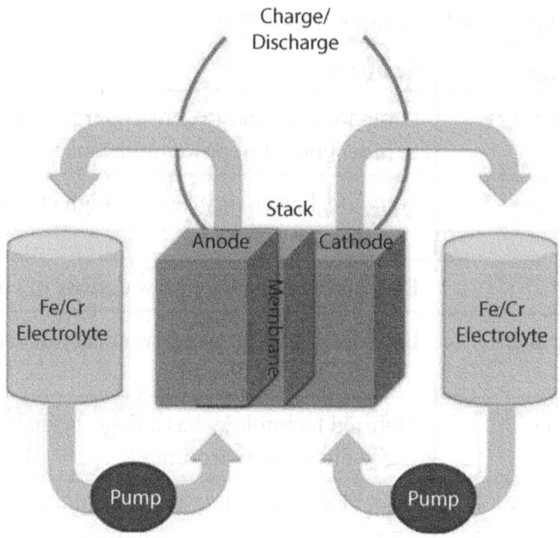

Figure 1.1 Representation of schematic for redox flow batteries.

cost-effective. Conventional rechargeable batteries lead to a simple and effective means for the storage of electricity. However, the development of such batteries focused on their transportation, portability, and size-to-volume ratio, which are some critical factors for grid storage batteries than in transportable applications. Thus, battery performance optimization may deliver superior results to reduce cost. This type of battery requires some features like durability for better charge/discharge cycles, better efficiency, good lifetime, and best output, and must be cost-effective. Table 1.1 compares the features of different energy storage systems.

Specifically, after 2010, creative outlines and workings of RFBs such as the utilization of flexible storage vessels along with flowable redox active resources magnetized attention for the growth of a variety of novel hybrid storage of energy and conversion systems. In conventional RFBs, salts of metals in an aqueous medium are employed as electrolytes and the disadvantage of these aqueous electrolytes is that the output cell voltages are intrinsically low (i.e., <1.7 V) due to water electrolysis issue. Therefore, efforts have been shifted toward the expansion of new redox flow chemistries [6]. VRFBs emerge as the most focussed RFB where the same metal cation is present in catholyte as well as in anolyte solutions. The transverse of vanadium ions via membrane is a reversible regeneration process that delivers a longer life to the solution of electrolytes. RFBs can be classified based on solvents (aqueous and non-aqueous) or species. Non-aqueous solvents are preferable for improving energy density, which indeed becomes essential for cells to gain a high voltage [7, 8]. Non-aqueous redox-active

Table 1.1 Comparative features for different energy storage systems.

Energy storage	Features
Redox Flow Battery	Long life, low energy density, large scale, independent of power and capacity
Super Capacitor	Long life, low energy density, high efficacy, short discharge time
Flywheel	Expensive, high power, low density of energy
Superconducting Magnets	Expensive, high power, low energy density
Pumped Water	Matured technology has a huge capacity, limited by the physical environment
Lead-acid Battery	High recovery cost, low-cost initial investment, and short life span cause pollution
Compressed Air	Huge capacity, limited by the physical environment
Lithium-ion Battery	High probable safety threats, high density of energy, high power density
Metal-air Battery	High density of energy, poor charging/discharging performance
Sodium-sulfur Battery	Expensive, high energy density, high power density, poor safety

molecules notably widen the scope to select active materials. The main advantages of utilizing redox-active molecules are their sustainable nature, structural diversity, and tunability which is considered to be a vital feature of any material to increase the electrochemical stability and attune the RFBs' redox potential through molecular engineering strategies [9].

Irrespective of the advantages present in VRFBs, such as simple redox reactions, long life, etc., the applications continued to be limited. A correlated shortcoming of VRFBs is the reliability issue that results from the crossover of active species via IEM, which necessitates the periodic regeneration of the electrolytes [10]. Cation exchange membranes (CEMs) have pervading capacity of vanadium ions, as the membranes are well permeable for cations and protons. The membranes in RFBs separate the electrodes' compartments while permeating the ion transport and hence, it completes the circuit. The prime characteristics of membranes are chemical stability, mechanical stability, high ionic conductivity, and refinement to lessen the crossover of redox-active species, as shown in Figure 1.2 [11].

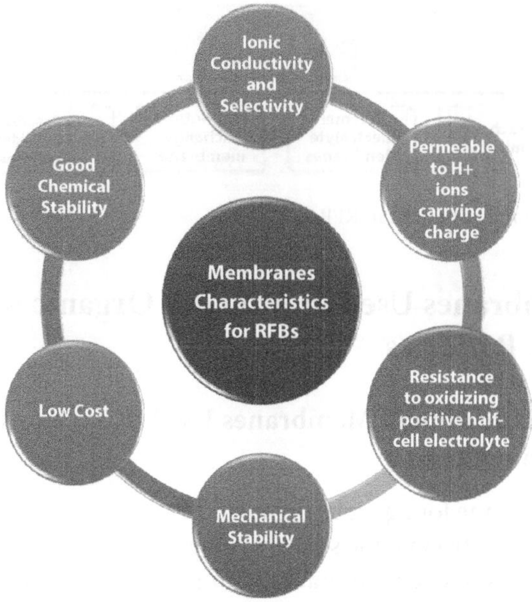

Figure 1.2 Characteristics of an ideal membrane for RFB applications.

In general, the selectivity contradicts the ionic conductivities, i.e., they are inversely proportional. So, depending upon the requirements of RFBs, these two characteristics should be balanced. Moreover, the membranes should be of a lower cost, directly improving the potential commercialization of RFB technology [12]. Most of the RFBs employ a per-fluorinated polymer membrane, Nafion, due to its capacity to resist strong electrolytes and vanadium ions of a high oxidation state (V^{+5}). But Nafion-based membranes are responsible for approximately 40% of cell stack charge [13]; thus, the necessity for RFB systems with less corrosive factors can reduce the cost. There are diverse variables engaged to characterize the RFB membranes, for example, ionic conductivity, permeability, selectivity, swelling capability, stability, etc. There present several membranes that are being widely utilized in RFBs, as shown in Figure 1.3.

In the present scenario, more focus has been given to the dimensions of research via more funding for flow batteries, especially for the research-oriented towards innovative redox couples along with creative flow-type systems. Therefore, the aforementioned chapter includes a brief outline of RFB membranes along with the working mechanism, chemistry involved in the system, etc. This focused chapter is not an extensive historic study of this area, but rather adequate to the current signs of progress in membrane technology used in RFBs.

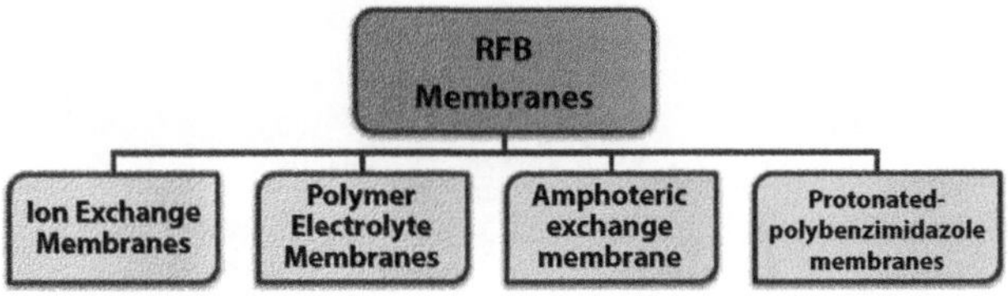

Figure 1.3 Membranes utilized in RFBs.

1.2 Membranes Used in Aqueous Organic Redox Flow Batteries

1.2.1 Classification of Membranes Used in Aqueous Organic RFBs

An ideal membrane for aqueous organic RFBs must satisfy the parameters discussed above. Strong efforts have been given by researchers across the globe to design suitable membranes with high efficiency and at low cost. Membranes used in Aqueous organic RFBs can be classified into the following types [14]. Nafion-based membranes

- Microporous membranes
- Anion-exchange membranes
- Cation-exchange membranes

1.2.1.1 Nafion-Based Membranes

Nafion-based membranes are the most extensively used membranes in Aqueous organic RFBs. Thomson *et al.* [15] reported the study of the performance of three compositionally same membranes (N115, NR-212, and NR-211) with thicknesses of 125 μm, 50 μm, and 25 μm respectively in vanadium RFBs. As shown in Figure 1.4, the coulombic efficiencies were estimated to be in the region of 97-98% and remain constant over 20 charge/discharge cycles. The energy efficiencies of Nafion N115 were ~75%, which was significantly lower than Nafion NR-211 and NR-212, ~82%. This illustrated the disadvantages of utilizing the thicker membrane.

Therefore, efficient electrical performance can be obtained by the utilization of thinner Nafion NR-211 and NR-212 membranes. Another advantage of using thinner membranes is that it could reduce the material cost by up to 12-15%.

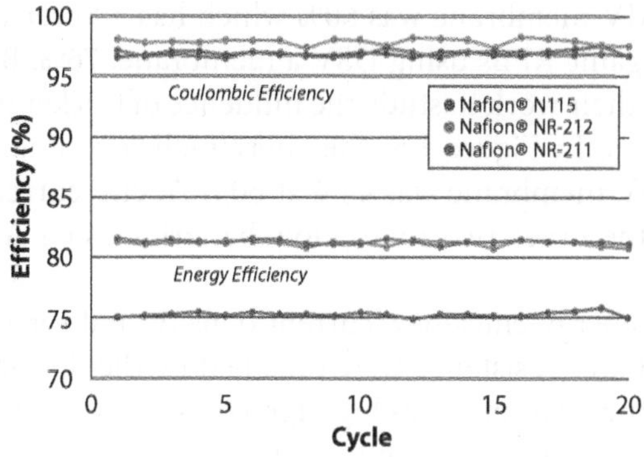

Figure 1.4 Coulombic and efficiencies of energy (in %) v/s the cycles for utilizing three Nafion membranes at current densities of 160 mA.cm^{-2} and at a flow rate of 1.2 L.min^{-1}. Reproduced with permission [15].

1.2.1.2 Microporous Membranes

Microporous membranes can have the potential to be alternative options to expensive membranes like Nafion membranes, Ion exchange membranes, etc., for RFBs. Janoschka *et al.* [16] reported a cheaper microporous dialysis membrane for a polymer-based aqueous redox flow battery. The membrane showed higher cycling stability even after 10,000 cycles and current density fit for 100 mA. cm^{-2}. The exhibited RFB indicated a promising economical RFBs by integrating basic membranes for dialysis, which are cost-effective with 5 to 10% of the cost of membranes based on Nafion, and also, in contrast to conventional IEMs, microporous membrane is substantially less expensive, making it an effective option for producing low-cost, highly effective membranes for aqueous organic RFBs [17].

1.2.1.3 Anion-Exchange Membranes (AEMs)

These membranes are extensively utilized in aqueous organic RFBs. AEMs have a remarkable influence on energy efficiency and the current density of RFBs. The impact of AEMs on the electrical performance of neutral aqueous organic RFBs was systematically studied and reported by B. Hu *et al.* [18] in 2017. The IEMs used were Selemion ASV, AMV, and DSV with different thicknesses of 120, 110, and 90 μm respectively. The reported data showed that the coulombic efficiencies of Aqueous Organic RFBs using DSV and AMV as membranes with NaCl as a supporting electrolyte was approximately 100%. The Voltage efficiency of Aqueous Organic

RFBs using AMV membrane was 60% which has a lower value than that of Aqueous Organic RFBs using DSV a membrane, 76%. B. Hu *et al.* further continued their work to study the influence of thicker membrane-like ASV membranes. In Figure 1.5, the energy efficacy of aqueous organic RFBs using ASV membrane was 44% at 60 mA. cm^2 current density that is lower than the other two redox flow batteries using AMV and DSV membranes.

The highest energy efficiency, current density, and capacity utilization with the lowest area resistance were obtained by the thinnest DSV membrane. Membrane resistance was highlighted as an essential component in determining RFB performance in this study.

1.2.1.4 Cation Exchange Membranes (CEMs)

In aqueous organic RFBs, CEMs are also often utilized. Orita *et al.*, in 2016, demonstrated the cycling performance of the CEM in TEMPOL, i.e., in Zn/4-hydroxy-2, 2, 6, 6-tetra-methyl piperidine-1-oxyl, depicted in Figure 1.6. CEM and sodium perchlorate are used as supporting electrolytes that exhibited a high retention capacity, i.e., 51% after the completion of 50 cycles, in comparison to various other combinations studied. Their findings also suggested that by altering the pH and removing impurities from the electrolyte, they may improve membrane function even more.

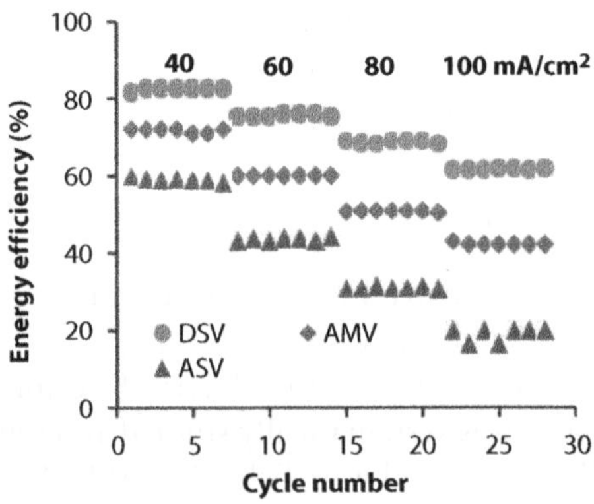

Figure 1.5 Energy efficiency (in %) against cycle number of FcNCl/MV aqueous organic RFBs considering three membranes, viz. DSV, AMV, ASV and 2.0 M NaCl were used as supporting electrolytes. Reproduced with permission [18].

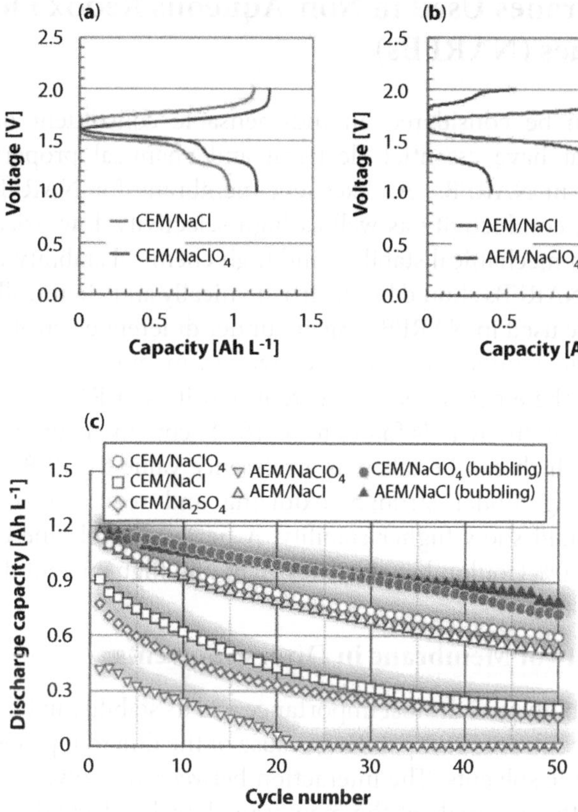

Figure 1.6 0.1 M supporting electrolyte was used for cycling of Zn/TEMPOL RFBs. First cycle voltage curve from constant current charge/discharge of 10 mA.cm^{-2} using (a) Cation exchange membranes, and (b) Anion exchange membranes; (c) up to 50 cycles of discharge capacity. Reproduced with permission [19].

Another application of CEMs on aqueous organic RFB reported by J. Winsberg *et al.* [20] in 2017, was an aqueous form of Zn/2, 2, 6, 6 tetramethyl piperidine-*N*-oxyl with a fumasep® F-930-RFD CEMs. AEMs were not used because they lost their ionic conductivity in the presence of Zn^{+2} (aq.) because of precipitation in zinc electrolytes. The presence of negatively charged sulfonic groups on the fumasep® F-930-RFD cation exchange membranes helps the retention of the active material TEMPO-4-sulfate in the cathode compartment, because of the repulsion between the sulfate group of active material and sulfonic groups on the membrane.

1.3 Membranes Used in Non-Aqueous Redox Flow Batteries (NARFBs)

Membranes can be considered an indispensable constituent of RFBs. Membranes that have essential electrical and chemical properties are very important in NARFBs. An excellent membrane for NARFBs must have high ionic conductivity as well as high selectivity, low price, lower swellability high mechanical stability, and high chemical stability [21, 22]. Membranes in NARFBs should be electrochemically and chemically stable in the electrolyte used in NARFBs. Again, under different electrolytic conditions, the ionic conductivity and selectivity of membranes vary, which has impacts on the efficiencies and current densities of RFBs. In lithium batteries, ionic conductivity is found to be 10^{-2} S. cm^{-1} in an ambient environment [23], which can be taken as a reference value for RFBs. During long-term cycling, under strongly oxidizing and reducing conditions membranes should show higher stability. A perfect membrane can help RFBs achieve 100% Coulombic efficiency (CE) and high current density.

1.3.1 Stability of Membrane in Diverse Solvents

Membrane's stability is of utmost importance and its stability in both aqueous and organic solvents vastly differ because of the different properties of water and organic solvents. The interaction between the solvent's organic molecules and the materials of the membrane largely affects the stability. Membranes like polymer membranes and membranes manufactured by using two-dimensional material get swollen when interacting with organic solvents. The amount of swelling ratio gets strongly influenced by the interaction between solvent molecules and membrane materials. Therefore, the reasons for the higher swelling rate can be attributed to the stronger interaction. Hence, stronger interaction can decompose the membranes. Table 1.2 corresponds to the comparison of different physical parameters of organic solvents that can be utilized in NARFBs.

The property permittivity can give information about the polarity of the solvents. As can be seen in the above Table 1.2, the solvents such as ethylene carbonate, dimethyl sulfoxide, propylene carbonate, acetonitrile, and dimethylformamide have high permittivity values (which means high polarity). Therefore, polar components of the membranes are easily dissolved by these solvents, leading to instability of the NARFBs. Again, solvents like ethyl methyl carbonate, diethyl carbonate, and dimethyl carbonate have low permittivity values (less polar solvents), and due to this, the less polar components on the membranes are easily dissolved.

Table 1.2 Different physical parameters of organic solvents utilized in NARFBs [24].

Solvents	Dipole moment (D)	Viscosity (mPa s)	Permittivity ($C^2N^{-1}M^{-2}$)	Density ($g\ cm^{-3}$)
Ethyl acetate	1.82	0.43	6.02	0.902
Ethylene carbonate	4.90	1.90	89.80	1.321
Ethyl methyl carbonate	0.51	0.65	2.93	1.006
Acetonitrile	3.53	0.34	35.90	0.776
Dioxolane	1.50	0.59	7.13	1.060
Dichloroethane	1.86	0.73	10.37	1.253
Dimethyl carbonate	0.93	0.58	3.13	1.069
Dichloromethane	1.55	0.39	8.93	1.327
Methyl acetate	1.72	0.36	6.68	0.932
Tetrahydrofuran	1.75	0.46	7.58	0.888
Dimethylformamide	3.24	0.80	36.70	0.948
Propylene carbonate	4.94	2.53	64.92	1.205
Diethyl carbonate	1.07	0.75	2.84	0.975
Dimethyl sulfoxide	4.06	1.99	46.50	1.100
Tetraethylene glycol dimethyl ether	1.92	4.01	7.68	1.009
Dimethoxyethane	1.71	0.46	7.20	0.868

1.3.2 Ionic Permeability and Selectivity

Due to the self-discharge property of the battery, the membrane must show non-selective diffusion across the membranes. Ionic selectivity can be defined by the ratio of the supporting electrolyte's permeability to that of the active species' permeability [25].

$$Ionic\ Selectivity = \frac{Permeability\ of\ the\ supporting\ electrolyte}{Permeability\ of\ the\ active\ species}$$

In general, the supporting electrolytes and active species permeabilities can be determined separately using an H-type diffusion cell, where one part is filled with supporting electrolytes whereas the other part is filled with active species. To avoid a concentration gradient, the solution in both half cells is vigorously stirred. UV-vis spectrophotometer is generally employed to monitor the concentration of active species. Fick's diffusion law is utilized to calculate the permeability shown as follows:

$$V\frac{dCt}{dt} = A\frac{P}{L}(C_o - C_t)$$

where V corresponds to the volume of the active species part, C_t corresponds to the active species concentration on the same side at time t, P corresponds to the membrane's permeability, A represents the membrane's effective area, L corresponds to the thickness, and C_0 represents the active species initial concentration in the other half-cell.

1.3.3 Ionic Conductivity

This can be observed by Electrochemical Impedance Spectroscopy (EIS) employing an H-type electrochemical cell [26]. Before all the measurements, the membrane is soaked for 24 hours in the electrolyte. The conductivity cell's resistance with and without membrane was measured over frequencies ranging from 1 MHz to 1 Hz [25]. By using the following equation, the ionic conductivity can be evaluated.

$$\sigma(S.cm^{-1}) = \frac{d}{R}$$

$$R\,(\Omega.cm^2) = A(r_1 - r_2)$$

where σ is the ionic conductivity, d corresponds to the thickness of the membrane, the effective area is represented by A, whereas R represents the resistance of the membrane, r_1 and r_2 correspond to the electrical resistance of conductivity cell with and without membrane, respectively.

1.3.4 Swelling

The swelling ratio is evaluated at room temperature by comparing the dimension of the membrane, before immersion and after immersion in the respective electrolyte for 24 h [27].

$$Swelling\ Ratio\,(\%) = \frac{Lw - Ld}{Lw} \times 100$$

where lengths of the membrane before immersion in the electrolyte (i.e., dried membrane) is represented by L_d and after immersion in the electrolyte (i.e., electrolyte saturated membrane) is L_w.

However, membrane swelling does not necessarily follow an isotropic pattern. The swelling ratio can also be calculated by comparing the changes in membrane area and width.

To get a better view of the compatibility of the electrolyte and membrane, electrolyte uptake emerges as an important factor. To calculate electrolyte uptake, the membrane is weighed before and after immersion into the respective electrolyte. After immersion, the adsorbent paper was utilized to remove the excess electrolytes from the membranes and after that, the weight of the saturated membrane is measured [28].

$$Electrolyte\ uptake\,(\%) = \frac{W_s - W_d}{W_d} \times 100$$

where W_d and W_s correspond to the weight of the membrane before and after being immersed in the electrolyte, respectively.

1.3.5 Mechanical and Chemical Stability

Mechanical stability of the membrane is obtained by calculating tensile strength, at ambient temperature using a universal tensile test machine, H5K-T [27]. The membrane's chemical stability can be evaluated by comparing the data obtained by XRD pattern, FT-IR, FESEM, and XPS characterization before and after a long-term cycling experiment.

1.3.6 Cycling Performance

The membranes' cycling performance is measured using a battery assembly. The factors like rates, efficiencies, retentions of the capacities, and the time of testing of NARFBs are all reliant on the membrane. The coulombic efficiency (CE) can be defined as the degree of the active species crossover and expressed as the ratio of charge to discharge capacities (equation 1.1). Voltage efficiency (VE) can be defined as the resistance provided by ion transfer via the membranes and can be delineated by the ratio of voltages of charge and discharge as shown in equation 1.2. And the efficiency of energy (EE) can be explained via the product of VE and CE, as shown in equation 1.3 [29].

$$CE\,(\%) = \frac{discharge\ capacity}{charge\ capacity} \times 100 \tag{1.1}$$

$$VE\,(\%) = \frac{discharge\ Voltage}{charge\ voltage} \times 100 \tag{1.2}$$

$$EE = CE \times VE \tag{1.3}$$

Unwanted diffusion of active species through the membrane during the cycling of NARFBs is called crossover, which leads to a decrease in the CE and causes unwanted self-discharge. Therefore, to maintain the capacity of the membrane, a perfect membrane must have low permeability of active species [30].

1.3.7 Classification of Membranes Used in NARFBs

Many efforts have been dedicated to enhancing the performance of membranes to fulfill the above-mentioned parameters. The membranes used in NARFBs are categorized into two types:

- Dense membranes
- Porous membranes

1.3.7.1 Dense Membranes

These membranes are composed of dense materials. Ion exchange membranes (IEMs) and dense ceramic membranes are examples of dense membranes. Ion exchange membranes show higher ionic selectivity and strong mechanical strength. These membranes consist of organic polymer with many ionic counterparts. This ionic counterpart allows the transport of selectively one type of ions while canceling out the opposite types [31]. Different IEMs that are utilized in NARFBs are listed in Table 1.3 [24].

Table 1.3 Different commercially available ion exchange membranes utilized in NARFBs.

Ion exchange membranes	Commercial name	Permeability ($\times10^{-9}$ m2 s^{-1})	Conductivity (mS cm^{-1})	Supporting electrolyte	Active species	Solvent	Current density (mA cm^{-2})	CE (%)	VE (%)	EE (%)
AEMs	Neosepta AHA	8.5	0.21	$TEABF_4$	Fe $(BiPy)_3(BF_4)_2$/ Ni $(BiPy)_3(BF_4)_2$	PC	-	-	-	-
	Neosepta AFX	-	0.33	$TEABF_4$	Fe $(BiPy)_3(BF_4)_2$/ Ni $(BiPy)_3(BF_4)_2$	PC	-	-	-	-
	Neosepta AFN	-	0.18[aq]	$TEDABF_4$/ $TEDAPF_6$	V $(acac)_3$	MeCN	-	-	-	-
	AMI-7001	-	1.1[aq]	$TEABF_4$ $TEAPF_6$ $TEAPF_6$	V $(acac)_3$ TEMPO/BP DBB/DMBP	MeCN MeCN MeCN	0.14/0.014 0.5 1	~50 81 72	- 53 47	- 42 -
	Fumasep FAP-PK	-	0.29	$TEABF_4$	Fe $(BiPy)_3(BF_4)_2$/ Ni $(BiPy)_3(BF_4)_2$	MeCN	2	94	82	-
	Fumasep FAP-375-PP	-	15[aq]	$TEABF_4$	4-Oxo TEMPO/ Camphorquinone	PC	1	80.3	88.8	71.3
	Fumasep FAP-450	6.13 -	0.5 0.35	$TEABF_4$ $TBAPF_6$	Fe $(BiPy)_3(BF_4)_2$/ Ni $(BiPy)_3(BF_4)_2$ Cobalt(II) complexes	PC MeCN	2 0.0044	90 ~95	87 ~76	- ~72
CEMs	Nafion 117	88	10	$LiBF_4$ $LiBF_4$ $TBABF_4$	BCF3EPT/TMeQ DBBB/TMQ FcPI	PC PC DOL	0.14 0.0625 2 C[b]	~92 ~70 97.3	- - -	- ~47 -
	Nafion 1035	-	10.7	$TEABF_4$	$V(acac)_3$	MeCN	1	91	-	80
	Nafion 115	-	5.9	$TEABF_4$	$V(acac)_3$	MeCN	10	~95	~93	~88

(Continued)

Table 1.3 Different commercially available ion exchange membranes utilized in NARFBs. (*Continued*)

Ion exchange membranes	Commercial name	Permeability ($\times10^{-9}$ m2 s^{-1})	Conductivity (mS cm^{-1})	Supporting electrolyte	Active species	Solvent	Current density (mA cm^{-2})	CE (%)	VE (%)	VE (%)
	Nafion 212	77.8	3.75	$TEABF_4$	Fe $(BiPy)_3(BF_4)_2$/ Ni $(BiPy)_3(BF_4)_2$	PC	1	~80	~87	~69
	CMI-7000	-	1.5^{aq}	$TBAPF_6$	Fc + TEMPO/ $[Co(Cp)_2]PF_6$+MePh	MeCN	0.1	~94	-	-
	Fumapem F-14100	-	65^{aq}	$TEABF_4$	Fe $(bpy)_3(BF_4)_2$/Co $(bpy)_3(BF_4)_2$	PC	2.2	~87	~92	~81
	Nepem-117	-	58.7^{aq}	$NaClO_4$	TEMPO/MePh	MeCN	0.35	90	-	-

Note: aq represents aqueous solution; C^b represents the value which is not transformed to mA cm^{-2}; DBBB represents the 2,5 di tertiary butyl-1, 4-bis (2-methoxy ethoxy) benzene; Ferrocene (Fc); Tetrabutylammonium hexafluorophosphate ($TBAPF_6$); 2,3,6-trimethylquinoxaline (TMQ); Tetraethylammonium tetrafluoroborate ($TEABF_4$); Acetonitrile (MeCN); NMethylphthalimide (MePh); Fe (Bi Py)$_3$ $(BF_4)_2$ corresponds to Iron tris (2, 2′-bi pyridine) tetra-fluoro-borate; 4,40-dimethylbenzophenone (DMBP); 4-Oxo 2, 2, 6, 6-tetra-methyl-1-piperidinyl-oxy (4-Oxo TEMPO); Cobaltocenium hexa-fluoro-phosphate $[Co(Cp)_2]PF_6$; Propylene carbonate (PC); N-ferro-cenyl phthalimide (FcPI); Tris(2, 2′-bi pyridine) iron tetra-fluoro-borate (Fe $(bpy)_3$ $(BF_4)_2$); Tris(2, 2′-bi pyridine) cobalt tetrafluoroborate ($Co(bpy)_3(BF_4)_2$); 2,3,6-trimethylquinoxaline (TMeQ); BCF3EPT corresponds to 3, 7-bis (tri fluoro methyl)-N-ethyl phenothiazine; Benzophenone (BP); 2, 5-di tertiary-butyl-1, 4-di methoxy benzene (DBB); Ni (Bi-Py)$_3$ $(BF_4)_2$ corresponds to Nickel tris (2, 2′-bi pyridine) tetra-fluoro-borate; $TEDABF_4$ corresponds to 1-butyltriethylamine tetrafluoroborate; $TEDAPF_6$ corresponds to 1-buthyltriethamine hexafluorophosphate.

1.3.7.2 Dense Ceramic Membranes

Dense ceramic membranes are also widely used because of their good ionic selectivity and low ionic conductivity. For example, the Na-β-Al_2O_3 membrane was used by Muthuraman *et al.* [32] on vanadium-NARFBs. Electrochemical impedance spectroscopy data showed that the ionic conductivity was 2.97 x 10^{-2} S. cm^{-1} of the membrane, where $NaClO_4$ was used as the supporting electrolyte and acetonitrile as a solvent. The UV-visible study confirmed that the Na-β-Al_2O_3 membrane can lower the capacity loss because of the lack of crossover through it, during the cycling of vanadium NARFBs. But the membrane had only shown 11% current efficiency and 16% voltage efficiency and 15% state of charge. Zhang *et al.* [33] in 2018 reported a Lithium-based NARFB in which $Li_{1.5}Al_{0.5}Ge_{1.5}P_3O_{12}$ (LAGP) was used as a membrane. They used lithium bis(trifluoromethane)sulfonamide (LiTFSI) as a supporting electrolyte and dimethylformamide/dichloroethane as a solvent. During cycling, the battery shows 99% CE, and the capacity of the battery at 0.2 mA.cm^{-2} was maintained at approximately 90% after 50 cycles.

Even though dense ceramic membranes have good ionic selectivity, their applicability is limited by their low ionic conductivity, which leads to a lower value of current density, as well as their fragile nature and high cost [34].

1.3.7.3 Porous Membranes

Porous membranes usually consist of a solid matrix with pores of different sizes. Commercial porous membranes are grouped into two types: The dramatic porous membrane and the Celgard porous membrane. Wei *et al.* [14] in 2016 studied three Daramic and two Celgard porous membranes to investigate their effective application in the DBMMB (2,5-di-tert-butyl-1-methoxy-4-[2′-methoxyethoxy] benzene) redox flow battery where McPh (*N*-Methylphthalimide) as active materials and LiTFSI in DME solvent as supporting electrolyte. All the membranes used had different thicknesses and pore sizes. The Daramic membranes used are Daramic-175, Daramic-450, and Daramic-800 where 175 μm, 450 μm, and 800 μm are their thickness, respectively. The Celgard membranes used are Celgard-2325, and Celgard-4560 with 28 nm and 64 nm pore sizes, respectively. They reported the impact of different thicknesses and pore sizes on the resistance and crossover of the NARFBs. When the membrane thickness is reduced, the batteries area-specific resistivity reduces (Figure 1.7a), which is reflected in considerably enhanced VEs of respective NARFBs (Figure 1.7c). Again, membranes that have lower thicknesses yield high self-discharge rates and excess crossover of active material (as shown in Figure 1.7b); therefore, lower coulombic efficiencies at the same testing currents. Hence, the lower thickness of both Celgard membranes shows

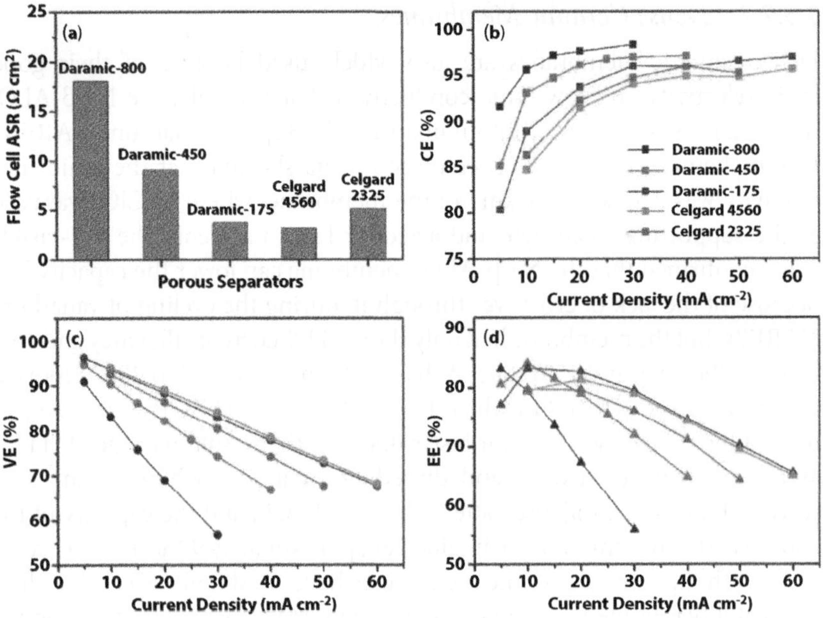

Figure 1.7 Electrochemical performance of NARFBs with various commercial porous membranes: (a) RFBs area specific resistivity before cycling; (b) coulombic efficiencies (CE); (c) voltage efficiency (VE); (d) energy efficiency (EE). Reproduced with permission [35].

lower coulombic efficiencies than Daramic membranes. On the other hand, Wei *et al.* [35] further reported that different pore sizes of porous membranes can have an effective influence on the membrane performance also. The large size of the pores of porous membranes shows relatively low resistance and high current flow in NARFBs and vice versa. For example, Celgard 2325 with 28 nm pore size shows high area-specific resistance (5.1 Ω. cm^2) and low VEs than the 7-fold thicker Daramic-175 membrane (Figures 1.7a and 1.7c). Among all the above-mentioned porous membranes, Daramic-175 shows the highest energy efficiencies and excellent performance in high current operations (Figure 1.7d). Table 1.4 represents the various commercial membranes applied for NARFBs [24].

1.4 Ion-Exchange Membranes or Ion-Conducting Membranes for RFBs

In RFBs, the most important component is the membrane. The membrane does not allow the electrolytes, positive/negative to mix in the two half

Table 1.4 The reported commercial porous membranes applied in NARFBs.

Commercial name	Pore size (μm)	Porosity (%)	Thickness (μm)	Permeability ($\times 10^{-6}$ m^2s^{-1})	Conductivity (mS cm^{-1})	Supporting electrolyte	Active species	Solvent	Current density (mA cm^{-2})	CE (%)	VE (%)	EE (%)
Celgard 4560	0.064	55	110	-	3.2 8.5	LiTFSI LiTFSI	DBMMB/MePh DBMMB/ BzNSN	DME MeCN	- 40	- ~90	- -	- ~60
Celgard 3501	0.064	55	25	-	-	LiTFSI	Fc1N112-TFSI/ Li	EC/ PC/ EMC	3.5	99	88	87
Celgard 2500	0.064	55	25	-	0.1	$TEABF_4$	Fe $(BiPy)_3(BF_4)_2$/ Ni $(BiPy)_3(BF_4)_2$	PC	-	-	-	-
Celgard 2400	0.043	42	25	- 2.43	1.0 -	$LiPF_6$ $TEABF_4$	Fe$(BiPy)_3$ $(BF_4)_2$/ Ni$(BiPy)_3$ $(BF_4)_2$ V$(acac)_3$	EC/ EMC MeCN	- 1	- 49	- -	- 32.4
Celgard 2325	0.028	39	25	0.48	0.49	LiTFSI	DBMMB/MePh	DME	-	-	-	-
Daramic 800	0.15	57	800	-	4.4	LiTFSI	DBMMB/MePh	DME	-	-	-	-
Daramic 450	0.15	57	450	-	~5	LiTFSI	DBMMB/MePh	DME	-	-	-	-
Daramic 250	0.15	57	250	-	6.1	TEATFSI	DBMMB/ BzNSN	MeCN	20	85.7	94	80.6
Daramic 175	0.15	57	175	-	4.6	LiTFSI	DBMMB/MePh	DME	35	90	77	69

Note: 2,1,3-benzothiadiazole (BzNSN); Lithium bis(trifluoromethane)sulfonimide (LiTFSI); Ferro-cenyl-methyl di-methyl ethyl ammonium bis (tri-fluoro-methane) sulfonimide (Fc1N112-TFSI); Ferro-cenyl-methyl di-methyl ethyl ammonium tetra-fluoro-borate (Fe$(bpy)_3(BF_4)_2$); Iron tris(2,2′-bypyridine) tetra-fluoro-borate (Fe$(BiPy)_3(BF_4)_2$); Tetra-ethyl-ammonium tetra-fluoro-borate ($TEABF_4$); Nickel tris(2, 2′-bipyridine) tetra-fluoro-borate (Ni$(BiPy)_3(BF_4)_2$); Ethylene-carbonate (EC); Ethyl-methyl-carbonate (EMC); 2, 5 di-tertiary butyl-1-methoxy-4-[2′-methoxy-ethoxy] benzene (DBMMB); *N*-Methyl-phthalimide (MePh).

cells, which otherwise would result in a short circuit of the two electrodes. The membrane then allows only the ions to transfer, thereby completing the circuit for effective current flow [36, 37]. Hence it has to be chosen with utmost effort to get the most suitable one, having characteristics like good mechanical properties, good chemical resistance, resistivity to oxidizing atmosphere provided due to the electrolyte, should possess a low electrical resistance, must give less access to the transfer of poly-halide ions and give high passage to charges with hydrogen ions and ultimately, cost-effective [38]. Alongside, the prevention of water through it is necessary so as not to allow dilution [39]. Figure 1.8 is the general representation of IEMs.

In 1890, Ostwald first developed the ion exchange membrane-dependent processes. He found that membranes if impermeable to cations/anions are

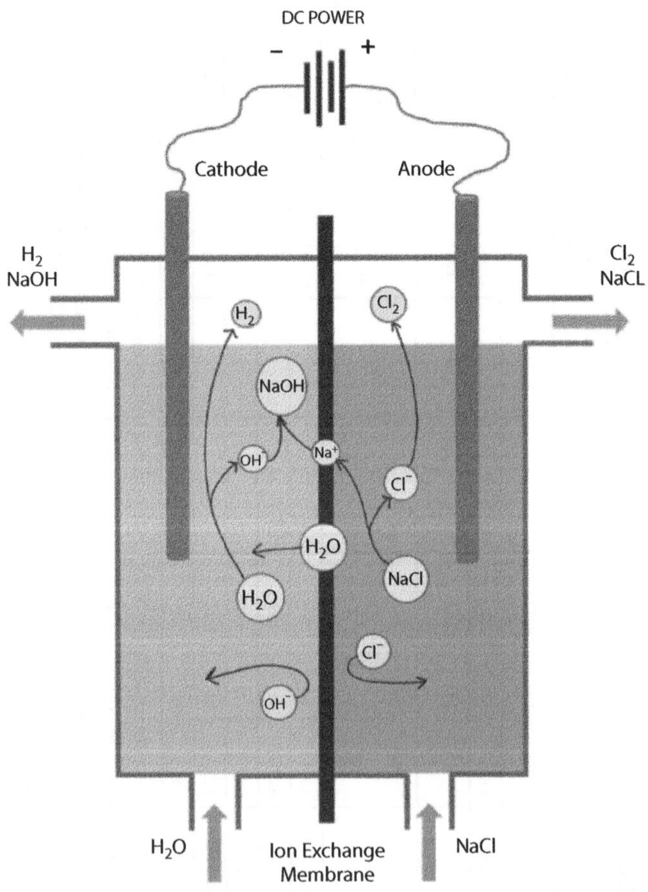

Figure 1.8 General representation of an IEM.

also impermeable to any electrolyte [40]. In 1939, membrane potential theory was proposed and then in 1949, a paper on bi-ionic potential was published by Sollner which deals with permeable selectivity [41, 42]. In 1950, a proposal was put forth stating that the IEMs could act as a separator in the electrolysis of NaCl solution and could produce gases of chlorine and hydrogen, and caustic soda [43]. These fluid-separating membranes are capable of ion exchange, which allows ions to move across the membrane along with water molecules. These membranes can be either sheets or tubes, or ribbons in shape [44]. Table 1.5 summarizes the characteristics of IEMs with cited examples.

Cross-linked linear polymer chains are used to design these membranes, a lack of which would otherwise cause the membrane to dissolve in water. IEMs have fixed ion groups. They also possess oppositely charged counter ions, to maintain a neutral exchanger system. These groups are known as the exchange sites which form the electrostatic bond with the oppositely charged ions [52]. During the exchange process, the ions are continuously

Table 1.5 General characteristics, applications, and cited examples for IEMs.

Characteristics	Application	Example	References
Ion conductivity	Electrodialysis	Concentration of electrolytes, ion exchange membrane processing alkali/acid, IE reactions through membrane, etc.	[45]
	Separator for electrolysis	Chlor-alkali production, organic synthesis by electrolysis, etc., synthesis of O_3, H_2O_2, etc.	[46]
	Battery	RFB, etc.	[47]
	Fuel cell	Hydrogen-oxygen, methanol-oxygen.	[48]
Hydrophilicity	Pervaporation	Pervaporation of organic solvents.	[49]
	Sensors	Gas sensor, medical	[50]
Fixed carrier (Ion exchange groups)	Gas separation	Removal/separation of gas, sugars, etc.	[51]

swapped using ions of a similar charge as the solution [53]. Figure 1.9 represents the scheme of IEMs.

Applications of IEMs range are wide, from the separation of ions to battery applications [54].

IEMs are broadly divided into five categories, viz. cation exchange (negatively charged), anion exchange (positively charged), amphoteric exchange, bipolar exchange, and mosaic exchange membranes.

In cationic exchange membranes, only the cations can pass through. In the anion exchange membranes, only the anions can permeate through selectively. For cations to pass through, the membrane should have cation exchange groups which can also be identified as anionic-charged groups. Similarly, for anions to pass through, the membrane needs to have anion exchange groups which are positively charged groups. At the amphoteric exchange membranes, there exist both exchange groups throughout the membrane at random, thereby allowing both cations and anions to pass through. Bipolar membranes have two membrane layers, one of cation exchange groups and another of anion exchange groups. This one is hence also known as a bilayer membrane. At the mosaic ion exchange membranes, cationic groups exist on the membranes along with anion exchange groups. This one is hence also known as a bilayer membrane. The cation exchange groups or the negatively charged groups utilized such as sulfonic acid, carboxylic acid, phosphonic acid, hydroxylic part of phenol group [55], thiol groups [56], *N*-oxide groups [57], and other groups with a negative charge. Anionic exchange groups or positive charge groups include amino groups, sulfonium groups, ammonium groups [58], phosphonium groups [59], etc.

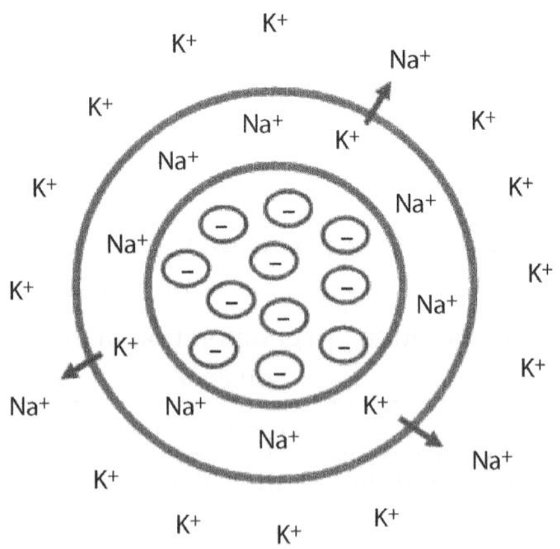

Figure 1.9 Scheme of an ion exchange process.

1.4.1 Cation Ion Exchange Membrane (CEMs)

Cation IEMs are commonly developed with copolymers and their ion exchange groups. Styrene is a copolymer and sulfonic acid groups (SIGs) act as ion exchange material in most hydrocarbon membrane cases. Here, SIGs are incorporated into the membranes by reacting them with sulfonating reagents [60]. In cation exchange membranes that have carboxylic acid groups, acrylic acid, methacrylic acid, and their esters, are copolymerized with divinylbenzene and/or other vinyl monomers. These copolymer membranes are then immersed in a hydroxide solution so that the ester groups of the membranes are hydrolyzed.

Phosphorus trichloride in the presence of a Lewis acid hydrolyses the phosphine dichloride groups with sodium hydroxide solution and then oxidizes the phosphonic acid to phosphonic acid to prepare the phosphorus-containing cation-exchange resins [61]. Grafting of vinyl-phenol or acetoxy vinyl-phenol on the polymer films is followed by hydrolysis of those groups to get a cation exchange membrane having phenolic groups [62].

1.4.2 Anion Exchange Membrane (AEMs)

Common AEMs are synthesized by the reaction of the amine with a co-polymer membrane. The AEMs have weakly basic anionic exchange groups, such as amino, that can be synthesized via the reaction of ammonia with a copolymer membrane, or amine, respectively. Polymer membranes having chloromethyl groups are generally prepared by copolymerization of chloro-methyl-styrene with other vinyl or/and divinyl monomers. These groups can be incorporated into aromatic groups in the presence of a Lewis acid via a reaction of chloro-methyl-methyl-ether [63]. For anionic membranes, the polymers are made of a moiety into which ion exchange groups can be easily incorporated and a part in which it is difficult to introduce. These two moieties then form hydrophilic and hydrophobic domains. After the polymer has been chloromethylated, a film is prepared by a casting method from this polymer. The film is then aminated with various amino compounds [64].

Membranes are further categorized by some properties like the nature of the materials, their pore size, and other morphological studies. Roughly, based on their microstructures, two categories are there for membranes, viz., homogeneous and heterogeneous membranes [65, 66]. These membranes are the most sought-after due to their good electrochemical properties and reproducibility [67].

Homogeneous membranes are prepared by different methods which are decribed below.

1.4.2.1 *Preparation by Condensation Reaction of Ionic Monomeric Compounds*

A low molecular weight viscous polymer is prepared via condensation reaction from the monomers and then coated on a fabric that is reinforced such as glass fiber and then cured for completion of the condensation reaction with drying and heating [68].

1.4.2.2 *Preparation by Polymerization of Vinyl Monomers*

A hydrocarbon ion exchange membrane is nothing but a co-polymer membrane mainly composed of two constituents, viz. styrene, and divinylbenzene, which have sulfonic acid groups or benzyl trimethylammonium groups. The ion exchange membrane with the starting material is a vinyl monomer; it is polymerized to prepare the ion exchange membrane, keeping in mind to maintain the homogeneous nature of the polymer if used more than one polymer [69]. After styrene has been partially polymerized, peroxide and additional styrene are added. The monomer and linear polymer mixture are polymerized and thin slices are made. Different exchange groups are then incorporated over the sliced films [70].

Another method involves the coating or pasting method where linear polymers without ion exchange groups are computed in a mixture of mono-vinyl monomers to get a paste-like consistency. This was then utilized as a coating material for the fabric of which the sides are covered using a separating film in which vinyl monomers are not impregnated [71]. This is then heated to get the film, which is then allowed to react with the reagents to introduce the ion exchange groups. This is a common method for commercial usage, resulting in good electrochemical properties and mechanical strength.

1.4.2.3 *Preparation from Conventional Polymers*

The grafting of vinyl monomers via a radiation-induced process on polymer films is a way to prepare ion exchange membranes and their precursors. Using an electron beam or γ-rays, the copolymers are irradiated. This in turn gives polymer radicals onto which the vinyl monomers are then embedded in the film. After that, the ion exchange groups are added into the imbedded copolymer to obtain the final product [72]. After a polymer film is allowed to swell in a mixture of vinyl monomers that comprises an initiator, it is enclosed through another film that does not swell under similar conditions and heated so that the monomers in the film polymerize. Ion exchange groups are later added to the resulting membrane [73, 74].

1.4.2.4 Preparation by Plasma Polymerization

Organic compounds such as per-fluorobenzene and SO_2, with plasma process, are polymerized, and resulting polymers are cast on a substrate to make perfluorocarbon cation exchange membranes [75].

Heterogeneous membranes, on the other hand, are prepared using an inert binder to the finely ground ion exchange membrane material. These bring out irregularities in the membrane itself, although these are easier to synthesize and have good mechanical properties. The heterogeneous membranes lack the electrochemical properties of the homogeneous ones. A suitable plasticizer is added at times to the membrane which is reinforced with woven fabric or polymers such as polyethylene or nylon. The polymer film containing the ion exchange material has a suspension of the ion exchange materials in an inert polymer solution which is deposited on a flat plate. After that, the solvent was evaporated to obtain the heterogeneous ion exchange membrane [76]. After preparing the ion exchange membranes, their properties and characteristics have to be evaluated. As ion exchange membranes have polar groups in a non-polar polymer matrix, they have a particular microstructure that allows these membranes to work more efficiently. They have several properties, viz. fixed ion concentration, water content, ion exchange capacity, transport number of counter-ions, electrical resistance (electrical conductivity), the diffusion coefficient of electrolytes, the diffusion coefficient of non-electrolytes, osmotic water transfer, electro-osmotic water transfer, Donnan adsorbed salts, permselectivity between ions of the same charge, stability toward chemicals, anti-oxidation properties, thermal stability, swelling degree, mechanical strength, etc. The most important properties are the transport number of the counter ions along with the electrical resistance that is provided by the membrane itself.

The electrical resistance of a membrane also varies with electrolyte solution concentration due to the increase in adsorbed salts and shrinking of the membranes with increasing external electrolyte solution [77].

Aqueous RFBs consists of two fully soluble redox couples. One of them gets electrochemically oxidized while the other gets reduced in the inert electrodes. After that, they are disconnected by an ion permselective membrane that can prevent cross-mixing. This whole procedure in turn generates electricity [78, 79]. Different sorts of RFBs include Fe-Cr redox battery systems with Fe anode solution and Cr cathode solution [80]. Here, hydrogen ions move via the membrane between the two electrolytes in the cycles of the battery [81].

If an anion exchange membrane is used, due to the high concentration of hydrochloric acid, chloro-complexes of Fe ions are formed and the membrane's electrical resistance increases due to the strong binding of the complexes to anion exchange group. Hence, a cationic exchange membrane with low electrical resistance and proton permeable selectivity is more desirable to decrease the internal resistance of the battery [82].

Of the cation exchange membrane redox flow batteries, the VRFB has garnered quite attention. It has almost 80% efficiency along with long cycle life [83, 84]. Most of the RFBs encountered cross-contamination of electrolytes. This was because of the different redox couple. Because vanadium ions exist as-V^{+2}, V^{+2}, VO^{+2}, and VO_2^+ in an aqueous solution, redox couples can be formed by all vanadium ions. Hence, the vanadium element in the half cells prevents cross-contamination.

1.5 Polymer Electrolyte Membranes

A flow battery consists of three different parts, viz. single cell, two electrolytic cells, and pump. Mainly, there are two different electrodes present in the cell, the positive and negative electrodes, along with a polymer electrolyte membrane (PEM). VRFBs have attracted tremendous interest due to their electrolyte, vanadium. The VRFB boosts the excellent operating cycle, with an average of 20,000 cycles [85]. The membrane in an RFB provides dual, ion exchange as well as prevention of cross-mixing. But the cost of developing the membrane, a polymer electrolyte one, is quite high. Hence attention has been given to cost-effectively developing them [86]. For a VRFB to be efficient, one of the most important factors is its high ionic conductivity. Batteries with a low ohmic loss are considered best for this purpose [87]. The membrane has to be super selective in differentiating between V^{+2}, VO^{+2}, V^{+3}, and VO_2^+ i.e., the variable states of vanadium ion [88, 89]. Good mechanical strength, chemical stability, and efficient water transport capability are a few of the other important properties required for am polymer electrolyte membrane to provide an effective redox flow battery. One of the most commonly used PEM is the Nafion series membranes. These membranes possess high chemical stability as well as ionic conductivity [90].

1.5.1 Membrane Properties

1.5.1.1 Ion Exchange Capacity

The main purpose of the membrane is to give access to the cations/anions at their exchange sites [91]. In the VRFB, this capacity is largely affected by the non-exchanged residual ions present in the electrolyte [92]. It is of utmost necessity that Ion Exchange Capacity is high for superior battery performance. According to the literature, the Nafion series has a capacity of 0.85-0.95 mmol. g^{-1} [93].

1.5.1.2 Chemical Stability

A membrane must resist the chemical effects in the battery system. Hence, they should possess excellent chemical stability so as not to corrode themselves. With a corroded membrane, the risk of cross-mixing of the electrolytes runs high, which in turn leads to short circuits and battery damage [94].

1.5.1.3 Thermal Stability

If the polymer membrane works even at a high temperature, the efficiency of the battery increases. It was reported that elevating the temperature range from 15 to 55 °C incrementally improved the battery's efficiency from 259.5 to 349.8 mW. cm^{-2} [95]. A VRFB normally works below 100 °C, and the thermogravimetric analysis shows that for the Nafion series the stability decreases tremendously after 400 °C [96].

1.5.1.4 Mechanical Property

Membranes with higher mechanical strength increase the battery life as they are less prone to damage [97].

1.5.1.5 Ionic Conductivity

Ionic conductivity means easier transport of ions through a membrane, which in turn helps the battery run more smoothly with high voltage efficiency. Lower conductivity means increased resistance and quicker discharge of the battery [98].

1.5.1.6 Vanadium Ion Permeability

For vanadium-based electrolytes, one method of transport included positively charged ions accumulated on the membrane's surface while the other includes cation accumulation inside proton-conducting channels [99]. The membrane must selectively distinguish between the different states of vanadium present and give passage to only the ones that completed the battery with an efficient output. A high rate of crossover of vanadium ions would result in self-discharge [100]. A low rate, on the other hand, would mean a limited ionic conductivity which again is not the desired quality. Thus, the membrane should achieve a balance in this regard.

1.5.1.7 Water or Electrolyte Uptake

Water uptake is the capability of absorbing water. In a polymer electrolyte membrane, this leads to ion pair dissociation in the exchange groups. Nafion membranes have a ratio of around 30% [101].

1.5.2 Transport Mechanisms

The VRFB while operating allows the passage of the vanadium ions, protons, and H_2O across a membrane. This process follows a complex mechanism leading to concentration and volume difference of electrolytes in the two half cells.

1.5.2.1 Proton Transport

Proton, a positively charged entity in the electrolyte, is generated from the dissociation of sulphuric acid. This moves through the cation exchange membrane while it completes the cell circuit [102]. In the Nafion membrane, a mechanism named Grotthus is followed for this transport. Here H-bonds are generated across the membrane in which the proton moves across the center of the membrane [103]. Another mechanism, viz. vehicular, suggests the transport by imparting motion of ion-pairs over the sidewall of the pores of membranes. In case of larger pore size, better transport takes place, but also the risk of vanadium ion permeability runs high.

1.5.2.2 Vanadium Ion Transport

In the cation exchange membrane, the positive vanadium ions get transported during the charging cycle, and vice versa. Similarly, at the anion exchange site, the negatively charged vanadium ions get transported. The electrolyte flow rate and the diffusion in the half-cells influence the way of transport of these ions through the membrane. The thickness of the membrane also governs this transport. The thinner or more degraded the membrane is, the more crossover of vanadium takes place, along with the risk of cross-mixing [104].

1.5.2.3 Water (H_2O) Transport

The volume of the electrolyte is immensely affected by the H_2O mobility. Cross-over of H_2O can cause disbalance in the two half-cells. While the transfer of H_2O, a phenomenon named electro-osmotic drag occurs. This lets more water transfer than vanadium ions. Higher transfer of H_2O leads to a higher flux. But the complete prevention of H_2O transport would not be possible. Hence a membrane is to be optimized for balanced transfer.

1.5.3 Membrane Preparation

As discussed in the previous section, membranes are classified into various categories.

1.5.3.1 Cation-Exchange Membrane (CEM)

In CEMs, the negatively charged groups help the transfer of cations. Dupont's Nafion membranes are usually utilized in CEMs for their high chemical stability along with ionic conductivity. This membrane is modified via various processes.

1.5.3.1.1 Addition with Other Polymers

Nafion coating on the polymeric substrate is an effective way of modifying the membrane as this improves mechanical strength. Lower use of Nafion also adds to the cost-effectiveness of the system. Nafion is sprayed onto the polymer surface and the unused solvent is evaporated [105]. Another method involves blending the Nafion resin with other polymers like PVDF and PTFE. This blending method controls the swelling behavior while increasing the ion selectivity [106, 107]. A different way to combine Nafion with polymer is to immerse the membrane in the polymer solution to obtain a composite membrane.

1.5.3.1.2 Addition with Organic/Inorganic Materials

The introduction of organic or inorganic molecules along the membrane helps in blocking the permeability. A common technique is the LBL (layer by layer) placement of materials onto the membrane, where vanadium ions permeability is reduced up to a certain level [108].

1.5.3.1.3 Sulfonated Poly(ether-ether-ketone) (SPEEK)

This polymer shows the advantage of producing low-cost VRFB with high thermal and chemical stability. SPEEK contains a sulphonic group of less acidity and a hydrophobic microstructure. This leads to a narrower ion channel and a lower cross-mixing rate [109].

1.5.3.1.4 Sulfonated Polyimide (SPI)

This is a non-fluorinated polymer, aromatic, with high mechanical and chemical stability. SPI-based membranes have been found to effectively block vanadium permeability due to their crosslinking ability [110].

1.5.3.1.5 Sulfonated Poly(fluorenyl-ether-ketone) (SPFEK)

This is regarded as a suitable replacement for the costlier Nafion membrane. SPFEK is lower in cost and allows low permeability to vanadium [111]. PTFE helps produce a composite membrane with SPEEK. The SPEEK is dissolved into suitable solvents and then cast into a PTFE-based membrane [112].

1.5.3.1.6 Poly(ether-sulfone) (PES)

PES is used in processes like filtration (ultra/nano) for membranes with high stable cycles and stability preparation [113].

1.5.3.1.7 Poly(vinylidene-fluoride) (PVDF)

PVDF polymer is hydrophobic with high chemical stability along with high thermal and mechanical stability as well but lacks ionic conductivity due to electrolyte influence in VRFB. Styrene and maleic anhydride grafting on the polymer are carried out via irradiation by electron-beam induced [114].

1.5.4 Anion-Exchange Membrane

The anionic exchange membrane consists of cationic group on the polymer backbone. The disadvantage of an anion exchange membrane is less chemical stability and low ionic conductivity.

1.5.4.1 Polysulfone (PSF)

Compared to the Nafion membrane, the poly-sulfone/poly(vinylidene fluoride) membrane fabricated was crosslinked by the formation of cations, which has well-balanced properties of mechanical stability as well as conductivity [115].

1.5.4.2 Poly(aryl-ether-ketone) (PAEK)

When PAEK is used for preparation in anionic membranes, it was able to provide better battery performance along with capacity retention capability but lacks high conductivity, which limits the voltage efficiency but restrains the efficiency in the energy of the battery [116].

1.5.5 Amphoteric Membranes

A combination of both the exchange membranes, viz anion as well as cation is explored in this membrane. PVDF-based membranes are mostly used in this type of preparation. Into the PVDF, *N, N*-dimethyl aminoethyl methacrylate, and sodium styrene sulfonate were grafted and then protonated giving an amphoteric-ion exchange membrane [117].

1.5.6 Porous Membrane

The porous membrane helps in distinguishing the protons from cations by their size [118].

1.5.7 Polybenzimidazole (PBI)

A heterocyclic polymer, which has high chemical along with thermal stability, decent mechanical strength, and a high glass transition temperature, thereby making it extremely stable in the corrosive environment of the battery [119]. These membranes are treated with strong acids to make them more proton-conducting. Chen *et al.*, 2019 [120] used pyridine and imidazole rings to form the proton transport channels, which enhances the proton transport across the membrane.

1.5.8 Polyacrylonitrile (PAN)

PAN is a commonly used material for redox flow battery membrane preparation. The pore size of PAN is tunable, and thus can be decreased, which leads to higher ion selectivity. Moreover, the coating of polymer of microporous nature on PAN for membrane fabrication is another way to prepare porous membranes [121].

Even after all these developments in the membrane field, a lot of challenges are still to be faced to maintain the optimum balance with all the requirements as discussed above. The vanadium ion permeability has an undesirable impact on the ion conductivity of the membrane, thereby affecting the battery life. Similarly, ion cross-mixing and low ionic conductivity lead to a shorter battery system life. Pore size tuning and layered composite membranes do show a way for more development. However, the focus needs to be drawn and a study needs to be done to obtain a porous membrane with the right ratio of conductivity and permeability.

1.6 Amphoteric Ion-Exchange Membranes

In 1932, Sollner and his group first presented the amphoteric ion-exchange membranes (AIEM) which include the cationic as well as anionic groups. One of the most promising vast-scale energy storage in the case of flow batteries is the vanadium redox flow battery (VRFB). The core part of VRFBs is the amphoteric ion-exchange membrane (AIEM) which can make the electrolytes disparate and conduct protons. The AIEM containing anions and cations results in high ion conductivity, high stability, low vanadium ion permeability, and cost-effectiveness. The selectivity of ions in AIEMs is significant for the crossover of vanadium ions with various valences, limiting the battery capacity [122].

To improve the ion conductivity in AIEMs, non-fluorinated polymers like sulfonated poly ether ketone (SPEEK), sulfonated poly(sulfone) (SPSF), and sulfonated polyimide (SPI) have been used [123-126]. AIEMs are a promising material for VRFB applications [127, 128] because of their higher conductivity value along with good chemical stability [129, 130]. The most broadly marketed RFB system is the VRFB only.

Wang *et al.* [131] in their work developed a novel AIEM based on VRFBs by blending ammonium polyphosphate (APP) and SPEEK. They introduced a large number of NH_4^+ ions into the SPEEK membrane. The addition of APP can enhance the permeability of vanadium ions along with increased proton conductivity. The NH_4^+ ions present in APP can penetrate the V ions using Donnan/Manning exclusion effect and ionic crossing networks. The presence of ionic bonds can significantly improve proton conductivity in the SPEEK membrane due to fast proton transport channels formed by H-bond and microphase separation resulting from the interaction between SPEEK and APP. The SPEEK/APP amphoteric membranes exhibit excellent performance than the pristine SPEEK and Nafion membranes with high voltage, coulombic efficiency, energy, and good stability. The synthesized SPEEK/APP amphiprotic membrane has high battery efficiency, selectivity, and good durability, which is a panned VRFB separator.

Liu *et al.* [132] in their work developed a practical way to address the crossover of electrolyte and consequential side reactions in redox flow batteries due to the use of bipolar redox-active molecules (BRMs). However, there are some difficulties in developing BRMs as finding redox-active moieties and reaching high cell voltage compared to already existing FBs. Their study is based on the strategies to design BRMs that can give high voltage resulting from the fused conjugation that can regulate the redox potential of the redox-active moieties. According to their study, the most widely used moieties for the construction of BRMs are ketone and quaternary *N*-moieties, leading to a better voltage gap and electrochemical stability. However, the symmetrical FBs can manifest a higher voltage of about 2.5 V and have high cycling stability.

Qian *et al.* [133] in their paper proposed a composite membrane for the VRFB based on SPEEK. This membrane was fabricated by a combination of several amounts of sulfonated poly (aryl ether sulfone) (SPAES) with chains of sulfo-alkylamine. The addition of SPAES can significantly enhance the formation of synergistic proton transport channels, leading to a balance between proton conductivity and permeability. However, this membrane shows high selectivity of about 28.8×10^3 S. min. cm^{-3} compared to other membranes such as Nafion-212, pristine SPEEK, and Nafion-117. This SPEEK/SPAES-15 membrane also has high energy and coulombic efficiencies resulting in low permeability, high selectivity, and proton conductivity. They maintain their good stability as well as durability even at a high density of 150 mA. cm^{-2}. This facile technique can lead to the form of a proton exchange high-performance membrane via regulating a synergistic sulfonic acid proton network.

However, vanadium-based RFBs are costly and toxic in their oxidized form. In this regard, iron-chromium redox flow battery (ICRFB) materials made up of Cr and Fe with low oxidation states are cost-effective and eco-friendly. ICRFBs are considered the type of RFBs that is one of the most efficient, cost-effective, and large-scale energy storage systems compared to all the vanadium and Zn-Br RFBs. It is a type of RFB where redox active species adopt significant chromium and iron chloride resource that can dissolve HCl. ICRFB contains some active substances like chromium (Cr^{+2}/ Cr^{+3}) and iron (Fe^{+2}/Fe^{+3}) that are not harmful to humans and the environment. Compared to VRFBs, ICRFB has relatively the same internal resistance as sulfuric acid as the H_2SO_4 has its conductivity, which can act as a supporting electrolyte.

There are two types of ICRFB analytes: mixed and unmixed. In the latter, the chemical equilibrium of Cr takes place between the inactive to active at elevated temperatures, which simultaneously enhances the rate of equilibrium. But ion selectivity with the help of IEM can be limited at elevated temperatures. The penetration rate of Fe and Cr ions become efficiently greater when the reactants are allowed to react at elevated temperature. Therefore, the mixed reactant electrolyte mode contains identical catholyte such as H^+, Cr^{+3}, Fe^{+2}, and Cl^- for long-term operation. On the electrodes of each half cell, these selective electrolytes are dispersed which are then circulated by the pump. Compared to unmixed reactant solutions, mixed reactant solutions have several advantages, i.e., in such cases, high perm selectivity of membranes is not required. It may be more cost-effective regarding chromite ore production compared to the production of chemicals in pure form, as severance of Fe and Cr would be redundant. Elementary commutations of mixed reactants are their high current efficacy for better voltage improvement. The better current and voltage efficiency depends on the properties of the membrane as well as the parameters of the operating system to accomplish maximum competence [134].

1.7 Protonated Polybenzimidazole (PBI) Membrane

For the production of large-scale energy and its storage, scientists are trying to develop technologies that can be renewable. To increase the performance of flow batteries, the main features are high conductivity and selectivity in a wide range of pH in ion-conducting membranes (ICMs). The lack of understanding of ion transport mechanisms limits the improvement of their performances. Among all the newly evolved VRFBs, Polybenzimidazole (PBI) membranes are considered an auspicious material because of their compact morphology, which can effectively hinder the vanadium crossover [135]. However, pristine PBI cannot conduct protons, but the N-atoms present in each imidazole ring can be protonated using H_2SO_4.[136, 137] The formed imidazolium ion can easily interact with the excess H_2SO_4 and water to promote the proton's conductivity via the Grotthuss mechanism [138].

Because of their unique structures, the PBI membranes are primarily used in alkaline and acidic applications such as alkaline Zn–Fe flow batteries, VFBs, etc. [139, 140]. PBIs are polymeric and can be easily complexed by strong acids to form acid-doped polymer electrolytes. These membranes are considered to be ion-solvating membranes. The shallow phase-separated structure in PBI membranes by imbibing electrolytes into the polymer matrix can easily enable their ions to conduct, unlike the IEMs. The selectivity and conductivity of such membrane can be enhanced by using adsorbed alkaline/acid solutions. Apart from this, acid-doped membranes also have higher resistivity than undoped ones [141-144]. The active ions and -OH ions can resist even at a high concentration of acidic or alkaline solutions as the Donnan effect of the PBI membranes are relatively strong. Thus, PBI membranes are considered to be amphoteric ion-exchange membranes than ion-solvating membranes.

Several groups have investigated the mechanism of acid-doped PBI membranes so far. Bouchet *et al.* [145] in their work indicated that regarding proton migration, acid anions play an essential role. Using these mechanical details, they studied three characteristic thermodynamic parameters: entropy, activation volume, and enthalpy based on isothermal and isobar conductivities. However, these parameters are solely dependent upon the pressure and temperature, which follow the thermodynamical predictions and lead to confirmation of proton transfer from one imide to another where anionic species participated.

At elevated temperatures, PBIs carry exceptional resistance to inorganic and alkalis as they can retain their chemical and mechanical properties even after exposure to inorganic acids and alkalis. To improve proton conductivity in PBIs, alkalis such as KOH, LiOH, and NaOH, and acids such as H_2SO_4, $HClO_4$, H_3PO_4, HCl, and HNO_3 have been used as doping agent [146]. The presence of robust and polarizable H-bonding on doping agents makes it a suitable material for high intrinsic proton conductivity.

Dai *et al.* [147], in their work, investigated PBI membranes in both acidic and alkaline media for the pilot plant test. The charge carrier in PBI membranes mainly depends upon the pH. At a high pH, these membranes are cation-conductive and anion-conductive at a low pH. To increase the ion conductivity, the proton or hydroxide conductivity in the membrane phase should be enriched as the mobility of protons and hydroxide ions is several times that of saline ions. However, dense PBI membranes are inconvenient to the OH^- and H^+ in alkaline and acidic solutions due to the strong Donnan effect. To overcome this disadvantage, one can use the acid treatment to increase the ion conductivity by the Grotthuss mechanism. Their studies revealed that acid treatment could eventually increase the ion transport number and promote ion transport effectively. According to their studies, acid treatment of PBI can relatively enhance the efficiency and performance of alkaline Zn-Fe flow batteries and VFBs compared to pristine PBI membranes.

The PBI polymers can form adducts that contain ether amine or amide groups of strong acidity/basicity. However, the proton conductivity of PBIs doped with acid mainly depends on the concentration and dopant. Out of all the acidic doping agents, one of the most widely used is phosphoric acid.

Wang and co-workers were the first to study the proton conductivity doped with acids and demonstrated that the mobility of phosphoric acid is more diminutive in PBIs than that of free phosphoric acid. They suggested the possible interaction between the PBI and phosphoric acid, and such types of interactions show better proton conductivity with high thermal stability without using any humidifier [148]. The immersion of PBIs into the aqueous phosphoric acid solution can effectively improve the proton conductivity. Further, it has been confirmed that doping with phosphoric acid at a high level can enhance the proton transfer from the acid to imino groups of PBI [149, 150].

Doping of PBIs with excess phosphoric acid can lead to a decrease in activation energy with an increase in the doping level. Ikhsan *et al.* developed a PBI membrane treated with phosphoric acid (PA) and KOH, showing conductivity up to 18 mS cm^{-1} and 56 mS. $cm^{-1,}$ respectively. Treatment of PBIs using H_2SO_4 can lead to the generation of chemical wastes. Thus, they have also investigated the treatment at various temperatures and concentrations of H_2SO_4 [151]. The core roles in high-temperature proton exchange membrane fuel cells (HT-PEMFCs) play by the proton exchange membranes (PEMs). The features required for high-performance polymer electrolytes at high temperatures and low humidity are high proton conductivity and stable mechanical properties. However, PBIs doped with phosphoric acid can be considered one of the most efficient materials for proton conduction. Figure 1.10 corresponds to the schematic illustration of PEMFC.

$$\text{Overall Reaction: } O_2 + 2\,H_2 \rightarrow 2\,H_2O + \text{Energy}$$

$$\text{Anodic Half Cell: } 2\,H_2 \rightarrow 4\,H^+ + 4e^-$$

$$\text{Cathodic Half Cell: } O_2 + 4\,H^+ + 4e^- \rightarrow 2\,H_2O$$

Ma *et al.* [152] show an increase in proton conductivity by setting both temperature and relative humidity constant, due to the high doping level of acid. The doping of PBI membranes with an excess amount of H_3PO_4 becomes the reason for the increase in proton conductivity. Before the maximum protonation, the imino groups of PBI membranes will be protonated by the phosphoric acid continuously at a lower doping concentration. The diffusion of the proton mainly depends on the protonated and non-protonated imino groups on the polymeric chain, resulting in low proton conductivity.

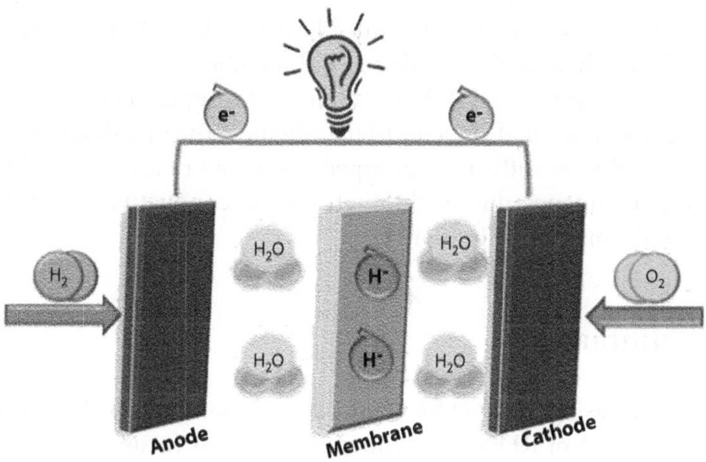

Figure 1.10 Schematic diagram of a PEMFC.

In addition to phosphoric acid-doped PBIs (Figure 1.11), sulfonated and sulfuric acid-doped PBIs have been developed, which are less dependent on doping concentration. For the synthesis of sulfonated PBI membranes, there are three possible mechanisms like polymerization [153, 154], chemical modification [155, 156], and thermal annealing [157]. However, the synthesis of sulfonated monomers is complicated among all these methods as they are not commercially available.

Wu and co-workers examined the interaction of PBI membranes and fuel cell performances with the doping level, H_2SO_4 concentration, temperature, and immersing time. The increase in the H_2SO_4 concentration can enhance the doping level in PBI membranes, which will lead to an increase in proton conductivity. The increase in proton conductivity results in outstanding performances of fuel cells. This doping method is different from phosphoric acid doping. This could be demonstrated based on the resistance of H_2SO_4 concentration in PBI membranes that increases in the catalyst layer with an increase in temperature [158].

Xu *et al.*, in their report, developed a sulfonated PBI (Figure 1.12), i.e., SOPBI (sulfonated poly [2,20 -(p-oxydiphenylene)- 5,50 -bibenzimidazole]) using post-sulfonated method. The post-sulfonation of PBI is one of the facile and practical methods compared to direct polymerization techniques with a controllable degree of sulfonation. The synthesized material shows good chemical and physical stability in DMSO with high mechanical and thermal stability, water stability, radical oxidative stability, and high proton conductivity. The degree of sulphonation emerges to be the main factor for the conductivity of SOPBI. The proton conductivity and performance of fuel cells of the H_2SO_4-loaded level in PBI membranes are essential, similar to phosphoric acid-doped PBI membranes [159].

The sulfuric acid-doped PBI membranes have higher proton conductivity than phosphoric acid-doped PBI membranes. The membrane electrode assembly (MEA) of sulfuric acid-doped PBI membranes exhibited higher fuel cell performance due to the high proton conductivity than Nafion under non-humid conditions [161].

1.8 Summary

The ever-increasing demand for energy storage on a large scale encourages the development of RFBs. Even though research is focused on RFBs, the progress of membranes is still limited due to the presence of several constraints. Membrane plays a key role in RFBs. Though many IEMs, mainly the fluorinated membranes viz., Nafion, are widely explored, mostly in VRFBs, they are still a long way from marketable employment owing to their expensive nature, high diffusivity of vanadium ions and transfer of water. As stated earlier, an ideal membrane must include high ionic conductivity and selectivity, superior chemical and mechanical stability, low-cost nature, resistance to oxidizing positive half-cell electrolytes, high permeability for protons, etc. Membranes are ubiquitous and are crucial in determining the efficiency of batteries, so enhancement in the technology of membranes is always imperative. The separator is an important module in batteries. The rich variety of suitable materials that can oblige as separators in RFBs permits innumerable paths for innovation. Overall, it can be noted

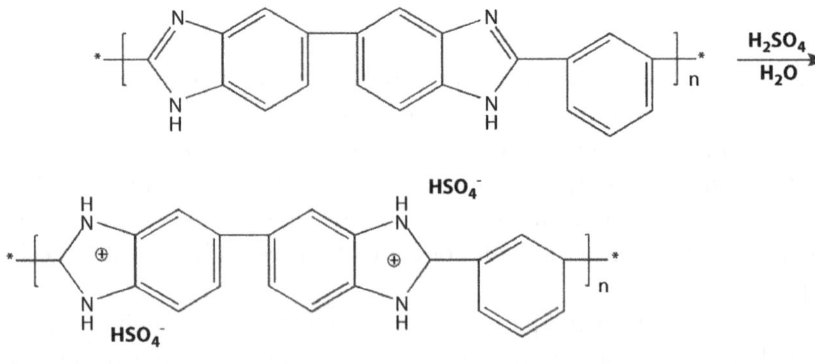

Figure 1.11 Acid doping of PBI.

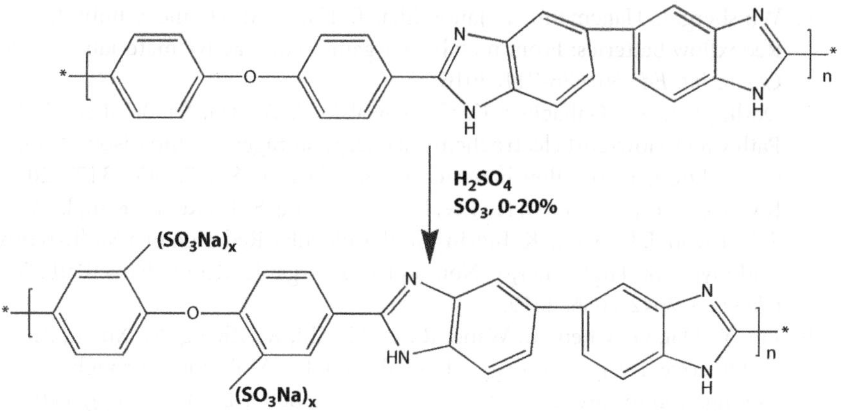

Figure 1.12 Post-sulfonation treatment of poly [2, 20 -(p-oxydiphenylene)-5, 50 -bibenzimidazole] (OPBI) membranes into sulfonated poly [2, 20 -(p-oxydiphenylene)-5, 50 -bibenzimidazole] (SOPBI) [160].

that the technologies of RFBs depend on the expansion and enhancement of applied polymer membranes, which is extremely challenging without a grasp of structure-property relationships, and due to this, the multifaceted relations between the membranes, active species, and electrolytes must be examined to yield enhanced materials. In the last couple of years, polymer research has been exploited largely for application in the storage of energy, along with refining the existing technologies while emerging novel concepts in varied areas. Due to the versatility of polymer membranes, the scientific community can further push the boundaries of RFBs. Briefly, in the future, continuous advancement in membranes will make RFBs a more promising possibility for large-scale energy storage.

References

1. Thaller, L. H. *Electrically Rechargeable Redox Flow Cell* (1976).
2. Kangro, W. A process for storing electrical energy, Germany patent 914264, 1949.
3. Skyllas-Kazacos, M., Rychcik, M., Robins, R. G., Fane, A. G., and Green, M. A. New all-vanadium redox flow cell. *J. Electrochem. Soc.*, 133, 1057-1058, 1986.
4. Skyllas-Kazacos, M., Chakrabarti, M.H., Hajimolana, S. A., Mjalli F. S., Saleem, M. *J. Electrochem. Soc.*, 158, 55-79, 2011.
5. Wang, W., Luo, Q., Li, B., Wei, X., Li, L., Yang, Z. *Adv. Funct. Mater.*, 23, 970-986, 2013.

6. Winsberg, J., Hagemann, T., Janoschka, T., Hager, M. D., and Schubert, U. S. Redoxflow batteries: From metals to organic redox-active materials. *Angew. Chem. Int. Ed.*, 56, 686-711, 2016.
7. Darling, R. M.; Gallagher, K. G.; Kowalski, J. A.; Ha, S.; Brushett, F. R., Pathways to low-cost electrochemical energy storage: A comparison of aqueous and nonaqueous flow batteries. *Energy Environ. Sci.*, 7, 3459-3477, 2014.
8. Kwon, G., Lee, K., Lee, M. H., Lee, B., Lee, S., Jung, S.-K., Ku, K., Kim, J., Park, S. Y., Kwon, J. E., Kang, K. Bio-inspired Molecular Redesign of a Multi-redox Catholyte for High-Energy Non-aqueous Organic Redox Flow Batteries. *Chem.*, 5, 2642-2656, 2019.
9. Liu, Y., Dai, G., Chen, Y., Wang, R., Li, H., Shi, X., Zhang, X., Xu, Y., Zhao, Y. Effective Design Strategy of Small Bipolar Molecules through Fused Conjugation toward 2.5 V Based Redox Flow Batteries. *ACS Energy Lett.*, 7, 1274-1283, 2022.
10. Agar, E., Benjamin, A., Dennison, C. R., Chen, D., Hickner, M. A., Kumbur, E. C. *J. Power Sources*, 246, 767-774, 2014.
11. Wang, W., Luo, Q., Wei, X., Li, L., Yang, Z. Recent progress in redox flow battery research and development. *Adv. Funct. Mater.*, 23, 970-986, 2013.
12. Schwenzer, B. Membrane development for vanadium redox flow batteries. *ChemSusChem.*, 4, 1388-1406, 2011.
13. Kamath H., Rajagopalan, S., and Zwillenberg, M. Vanadium redox flow batteries: An in-depth analysis. EPRI technical update, Palo Alto, CA, 2007.
14. Chen, H., Cong, G., & Lu, Y. C., Recent progress in organic redox flow batteries: Active materials, electrolytes and membranes. *J. Energy Chem.*, 27, 1304-1325, 2018.
15. Reed, D., Thomsen, E., Wang, W., Nie, Z., Li, B., Wei, X., ... & Sprenkle, V. Performance of Nafion® N115, Nafion® NR-212, and Nafion® NR-211 in a 1 kW class all vanadium mixed acid redox flow battery. *J. Power Sources*, 285, 425-430, 2015.
16. Janoschka, T., Martin, N., Martin, U., Friebe, C., Morgenstern, S., Hiller, H., Schubert, U. S., An aqueous, polymer-based redox-flow battery using non-corrosive, safe, and low-cost materials. *Nature*, 527, 78-81, 2015.
17. Prifti, H., Parasuraman, A., Winardi, S., Lim, T. M., & Skyllas-Kazacos, M., Membranes for redox flow battery applications. *Membranes*, 2, 275-306, 2012.
18. Hu, B., Seefeldt, C., DeBruler, C., & Liu, T. L., Boosting the energy efficiency and power performance of neutral aqueous organic redox flow batteries. *J. Mater. Chem.*, 5, 22137-22145, 2017.
19. Orita, A., Verde, M. G., Sakai, M., & Meng, Y. S., The impact of pH on side reactions for aqueous redox flow batteries based on nitroxyl radical compounds. *J. Power Sources*, 321, 126-134, 2016.
20. Winsberg, J., Stolze, C., Schwenke, A., Muench, S., Hager, M. D., & Schubert, U. S., Aqueous 2, 2, 6, 6-tetramethylpiperidine-N-oxyl catholytes for a

high-capacity and high current density oxygen-insensitive hybrid-flow battery. *ACS Energy Lett.*, 2, 411-416, 2017.

21. Zhang, H., Zhang, H., Li, X., Mai, Z., & Zhang, J., Nanofiltration (NF) membranes: the next generation separators for all vanadium redox flow batteries (VRBs)? *Energy Environ. Sci.*, 4, 1676-1679, 2011.
22. Qiao, L., Zhang, H., Lu, W., Dai, Q., & Li, X., Advanced porous membranes with tunable morphology regulated by ionic strength of nonsolvent for flow battery. *ACS Appl. Mater. Interfaces*, 11, 24107-24113, 2019.
23. Kamaya, N., Homma, K., Yamakawa, Y., Hirayama, M., Kanno, R., Yonemura, M., Mitsui, A., A lithium superionic conductor. *Nat. Mater.*, 10, 682-686, 2011.
24. Yuan, J., Pan, Z. Z., Jin, Y., Qiu, Q., Zhang, C., Zhao, Y., & Li, Y., Membranes in non-aqueous redox flow battery: A review. *J. Power Sources*, 500, 229983, 2021.
25. Peng, S., Zhang, L., Zhang, C., Ding, Y., Guo, X., He, G., & Yu., Gradient-Distributed Metal–Organic Framework–Based Porous Membranes for Nonaqueous Redox Flow Batteries. *Adv. Energy Mater.*, 8, 1802533, 2018.
26. Shin, S. H., Kim, Y., Yun, S. H., Maurya, S., & Moon, S. H., Influence of membrane structure on the operating current densities of non-aqueous redox flow batteries: Organic–inorganic composite membranes based on a semi-interpenetrating polymer network. *J. Power Sources*, 296, 245-254, 2015.
27. Bang, H. S., Kim, D., Hwang, S. S., & Won, J., Surface-modified porous membranes with electrospun Nafion/PVA fibres for non-aqueous redox flow battery. *J. Membr. Sci.*, 186-194, 2016.
28. Mai, Z., Zhang, H., Li, X., Bi, C., & Dai, H. Sulfonated poly (tetramethydiphenyl ether ether ketone) membranes for vanadium redox flow battery application. *J. Power Sources*, 196, 482-487, 2011.
29. Saha, M. S., Li, R., & Sun, X., High loading and monodispersed Pt nanoparticles on multiwalled carbon nanotubes for high performance proton exchange membrane fuel cells. *J. Power Sources*, 177, 314-322, 2008.
30. Lin, K., & Chen, Q., Gerhardt MR *et al.*, Alkaline quinone flow battery. *Science*, 349, 1529-32, 2015.
31. Xu, T., Ion exchange membranes: State of their development and perspective. *J. Membr. Sci.*, 263, 1-29, 2005.
32. Muthuraman, G., Boyeol, L., & Il-Shik, M., Na-β-Alumina as a Separator in the Development of All-Vanadium Non-Aqueous Tubular Redox Flow Batteries: An Electrochemical and Charging-Discharging Examination Using a Prototype Tubular Redox Flow Cell. *J. Electrochem. Soc.*, 165, A1920, 2018.
33. Zhang, H., Niu, Z., Ding, Y., Zhang, L., Zhou, Y., Guo, X., Zhang, X., Zhao, Y., Yu., G., Highly Concentrated Phthalimide-Based Anolytes for Organic Redox Flow Batteries with Enhanced Reversibility. *Chem.*, 4, 2814-2825, 2018.

34. Chen, H., Cong, G., & Lu, Y. C., Recent progress in organic redox flow batteries: Active materials, electrolytes and membranes. *J. Energy Chem.*, 27(5), 1304-1325, 2018.
35. Wei, X., Duan, W., Huang, J., Zhang, L., Li, B., Reed, D., Xu, W., Sprenkle, V., Wang, W. A high-current, stable nonaqueous organic redox flow battery, *ACS Energy Lett.*, 1, 705-711, 2016.
36. Mohammadi, T., Skyllas-Kazacos, M., Characterisation of novel composite membrane for redox flow battery applications. *J. Membr. Sci.*, 98, 77, 1995.
37. Vafiadis, H.; Skyllas-Kazacos, M., Evaluation of membranes for the novel vanadium bromine redox flow cell. *J. Membr. Sci.*, 279, 394, 2006.
38. Skyllas-Kazacos, M., Milne, N.A., Kazacos, G.C., Membrane properties and behaviour in the Generation 2 Vanadium Bromide Redox Flow batteries. In *Proceedings of the 16th International Federation for Heat Treatment and Surface Engineering Congress, Brisbane, QLD, Australia*, 72-77, 2007.
39. Sukkar, T., Skyllas Kazacos, M., Modification of membranes using polyelectrolytes to improve water transfer properties in the vanadium redox battery. *J. Membr. Sci.*, 222, 249, 2003.
40. Xu, T., Ion exchange membranes: State of their development and perspective. *J. Membr. Sci.*, 263, 1, 2005.
41. Teorell, T., An attempt to formulate a quantitative theory of membrane permeability. *Proc. SOC. Exptl. Biol. Med.*, 33, 282, 1935.
42. Sollner, K., The origin of bi-ionic potentials across porous membranes of high ionic selectivity. I. The bi-ionic potential and the mechanism of its origin; the various factors that determine the sign and the magnitude of the bi-ionic potential; the simplest chains in which bi-ionic potentials arise; systems with critical ions of the same size of different absorbability. *J. Phys. Colloid Chem.*, 53, 1211, 1949.
43. Juda, W. and McRae, W.A., Ion-exchange materials and method of making and using the same, US Pat. 2, 636, 851.
44. Paterson, R., An Introduction to Ion Exchange, Heyden and Son, Ltd.: London, UK, 1970.
45. Walters, W.R., Wieser, D.W., and Marek, L.J., Concentration of radioactive aqueous wastes, Electromigration through ion-exchange membranes. *Ind. Eng. Chem.*, 47, 61-67, 1955.
46. Kobuchi, Y., Motomura, H., Noma, Y. and Hanada, F., Application of ion exchange membranes to the recovery of acids by diffusion dialysis. *J. Membr. Sci.*, 1986, 27, 173-179; Sata, T., New applications of ion exchange membrane. *Macromolecules*, ed. J. Kohovec, VSP, Utrecht, Netherlands, 458-461, 1992.
47. Akita, S., Iwahori, T., Energy storage, *Karyoku Genshiryoku Hatsuden (Thermal and Nuclear Power)*, 52, 1344, 2001.
48. Grubb, W.T., and Niedrach, L.W., Batteries with solid ion-exchange membrane electrolytes. II. Low-temperature hydrogen-oxygen fuel cells. *J Electrochem. Soc.*, 107, 131, 1969; Perry, M.L., and Fuller, T.F., Historical

perspective of fuel cell technology in the 20th century. *J. Electrochem. Soc.*, 149, 559, 2002.
49. Wenzlaff, A., Boddeker, K.W., and Hattenbach, K., Pervaporation of water-ethanol through ion exchange membranes. *J. Membr. Sci.*, 22, 333, 1985.
50. Skarstrom, C.W., and Kertzman, J., Process for separating fluids and apparatus, US Pat 3, 735, 558.
51. Koval, C.A., Spontarelli, T., Condensed phase facilitated transport of olefins through an ion exchange membrane. *J. Am. Chem. Soc.*, 110, 293, 1988.
52. Kunin, R., Myers, R.J., *Ion Exchange Resins*; John Wiley & Sons Inc: New York, NY, USA, 1950.
53. Dorfner, K., *Ion Exchangers: Properties and Applications*; Ann Arbor Science Publishers: Ann Arbor, MI, USA, 1972.
54. Jacobson, K.S., Drew, D.M., He, Z., Efficient salt removal in a continuously operated up flow microbial desalination cell with an air cathode. *Bioresour. Technol.*, 102, 376, 2011.
55. Fujiwara, H., Asano, K., Takahashi, A., and Sugishita, A., Ion exchange membranes. Jpn. Pat. JP 50-92877, JP 50-105598 (unexamined application), 1975.
56. Lezzi, A, Cobianco, S., and Reggero, A., Synthesis of thiol chelating resins and their adsorption properties toward heavy metal ions. *J. Polym. Sci., Polym. Chem. Ed.*, 32, 1877, 1994.
57. Holt, P., and Tamami, B., Relation between pH and viscosity of some poly (alkylvinyl1pyridine N-oxides) in aqueous solution, *Die Makromol. Chem.*, 155, 55, 1972.
58. Kondo, S., Murayama, T., Takeda. Y., Tsuda, K., Polymer-supported sulfonium salts in liquid-liquid two phase reactions. *Makromol. Chem., Rapid Commun.*, 9, 625, 1988.
59. Iizawa, T., Yamada, Y., Ogura , Y., and Saito, Y., Novel synthesis of polymeric phase transfer catalyst containing polar unit. *J. Polym. Sci., Polym. Chem. Ed.*, 32, 1994.
60. Takata, K., and Sata, T., Modification of transport properties of ion-exchange membranes. XIV Introduction of -S02Cl groups with C1SO3H-H2SO4 mixture into membrane and formation of acid-amide bonding between polyethyleneimines and -S02C1 groups, *J. Macromol. Sci. Pure Appl. Chem.*, 33, 789, 1996.
61. Sata, T., Yoshida, T., and Matsusalu, K., Transport properties of phosphonic acid and sulfonic acid cation exchange membranes. *J. Membr. Sci.*, 120, 101, 1996.
62. Machi, S., Matsui, Y., Sugishita, A., Sugou, T., Taniguchi, H., Asano, K., and Fujiwara, H., Preparation method of graft polymers having fluorocarbon polymer backbone. Jpn. Pat. JP 57-18530 (examined application)
63. Sheng, Q., and Stover, H.D.H., A new route to (chloromethyl)styrene polymers, *Macromolecules*, 30, 6712, 1997.

64. Sugaya, Y., Terada, I., and Sanekata, K., New anion exchanger with cross-linking structure, Jpn. Pat. No. 2522356.
65. Strathmann, H., Synthetic membranes and their preparation. In *Proceedings of the NATO Advanced Study Institute on Synthetic Membranes. Science, Engineering, and Applications, Alcabideche, Portugal*, 1983.
66. Bungay, P.M., Lonsdale, H.K., de Pinho, M.N., Eds.; Reidel Publishing Company: Dordrecht, The Netherlands, 1986.
67. Choi, J.H., Kim, S.H., Moon, S.H., Heterogeneity of ion-exchange membranes: the effects of membrane heterogeneity on transport properties. *J. Colloid Interface Sci.*, 241, 120 2001.
68. Cadotte, J.E., King, R.S., Majerle, R.J., Pedersen, R.J., Interfacial synthesis in the preparation of reverse osmosis membranes. *J. Macromol. Sci. Chem.*, A15, 727, 1981.
69. Polymer Handbook, ed. J. Brandrup and E.H. Immergut, Interscience Publishers, New York.
70. Ishibashi, N., Seiyama, T., and Sakai, W., Preparation of cation exchange membrane and its electrochemical properties. *J. Electrochemi. Soc. Jpn.*, 22, 684, 1954.
71. Mizutani, Y., Yamane, R., Ihara, H., and Motomura, H., Studies on ion exchange membranes. XVI. The preparation of ion exchange membranes by the 'Paste Method'. *Bull. Chem. Soc. Jpn.*, 36, 361, 1963.
72. Hegazy, El.S.A., Dessouuki, A.M., Rabie, A.M., and Ishigaki, I., Membranes obtained by preirradiation grafting of acrylic acid onto poly(tetrafluoroethy-1ene-perfluoro-vinyl ether). *J. Polym. Sci., Polym. Chem. Ed.*, 22, 3673 1984.
73. Dimov, A.D., and Alexcandrova, I., Preparation of ion exchange membranes from modified polyethylene with high molecular weight. *Desalination*, 46, 361, 1983.
74. Hani, H., Hiraga, T., and Nishihara, R., Preparation method of ion selective membranes. Jpn. Pat. JP 35-13009.
75. Inagawa, N., Tasaka, S., and Horikawa, Y., Nafion-like thin film plasma-polymerize from perfluorobenzene/SO2 mixture. *J. Polym. Sci., Polym. Chem. Ed.*, 27, 3495, 1989.
76. K. Hayashi C. Kondo and J. Ueda, Preparation method of ion exchange membrane, *Jpn. Pat. JP* 39-19037.
77. Pismenskaya, N., Laktionov, E., Nikonenko, V., Attar, A.El., Auclair, B., and Pourcelly, G., Dependence of composition of anion-exchange membranes and their electrical resistance on concentration of sodium salts of carbonic and phosphonic acids. *J Membr. Sci.*, 181, 185, 2001.
78. Assink, R.A., Fouling mechanism of separator membranes for the iron chromium redox battery. *J Membr. Sci.*, 17, 205, 1984.
79. Kummer, J.T., and Oei, D.G., A chemically regenerative redox fuel cell. *II. J Appl. Electrochem.*, 15, 619, 1985.
80. Sata, T., Trends in studies and application of ion exchange membranes, *Soda Enso (Soda Chlorine)*, 38, 43, 1987.

81. Ohya, H., Emori, K., Ohne, T., Negishi, Y., and Matsumoto, K., Studied on membranes for a redox-flow battery 11. Several factors influencing the electrical resistivity of anion exchange membranes, *J. Electrochem. Soc. Jpn.*, 53, 462, 1985.
82. Ogumi, Z., Uchimoto, Y., Tsujikawa, M., and Takehara, Z., Modification of ion exchange membrane surface by plasma process. 1. H+ ion permselective membrane from Nafion for redox-flow battery, *J. Electrochem. Soc.*, 137, 1430, 1990.
83. Skyllas-Kazacos, M., Chakrabarti, M.H., Hajimolana, S.A., Mjalli, F.S., Saleem, M., Progress in flow battery research and development. *J. Electrochem. Soc.*, 158, 55, 2011.
84. Skyllas-Kazacos, M., Bankstown, R.M., Lindfield, R.R. AllVanadium Redox Battery. U.S. Patent 4, 786, 567, 1988.
85. Ulaganathan, M., Aravindan, V., Yan, Q., Madhavi, S., Skyllas-Kazacos, M., Lim, T.M., Recent advancements in all-vanadium redox flow batteries. *Adv. Mater. Interfaces*, 3, 2016.
86. Zeng, Y.K., Zhao, T.S., An, L., Zhou, X.L., Wei, L., A comparative study of all vanadium and iron-chromium redox flow batteries for large-scale energy storage. *J. Power Sources*, 300, 438, 2015.
87. Elgammal, R.A., Tang, Z., Sun, C.N., Lawton, J., Zawodzinski, T.A., Species uptake and mass transport in membranes for vanadium redox flow batteries. *Electrochim. Acta*, 237, 1, 2017.
88. Lu, W., Yuan, Z., Zhao, Y., Zhang, H., Zhang, H., Li, X., Porous membranes in secondary battery technologies. *Chem. Soc. Rev.*, 46, 2199, 2017.
89. Yang, Z., Zhang, J., Kintner-Meyer, M.C., Lu, X., Choi, D., Lemmon, J.P., Liu, J., Electrochemical energy storage for green grid. *Chem. Rev.*, 111, 3577, 2011.
90. Jiang, B., Wu, L., Yu, L., Qiu, X., Xi, J., A comparative study of Nafion series membranes for vanadium redox flow batteries. *J. Membr. Sci.*, 510, 18, 2016.
91. Kundu, P., Dutta, K., Kumar, P., Bharti, R., Kumar, V., Kundu, P., Polymer electrolyte membranes for microbial fuel cells: *Part A. Nafion-based membranes*. Amsterdam, The Netherlands: Elsevier; 2018.
92. Moukheiber, E., De Moor, G., Flandin, L., Bas, C., Investigation of ionomer structure through its dependence on ion exchange capacity (IEC). *J. Membr. Sci.*, 389, 294, 2012.
93. Trogadas, P., Pinot, E., Fuller, T.F., Composite, Solvent-Casted Nafion Membranes for Vanadium Redox Flow Batteries. *Electrochem. Solid-State Lett.*, 15, 2012.
94. Mohammadi, T., Kazacos, M.S., Evaluation of the chemical stability of some membranes in vanadium solution. *J. Appl. Electrochem.*, 27, 153, 1997.
95. Zhang, C., Zhao, T.S., Xu, Q., An, L., Zhao, G., Effects of operating temperature on the performance of vanadium redox flow batteries. *Appl. Energy*, 155, 349 2015.
96. Jana, K.K., Lue, S.J., Huang, A., Soesanto, J.F., Tung, K.L., Separator Membranes for High Energy-Density Batteries. *ChemBioEng Reviews*, 5, 346, 2018.

97. Teng, X., Dai, J., Su, J., Yin, G., Modification of Nafion membrane using fluorocarbon surfactant for all vanadium redox flow battery. *J. Membr. Sci.*, 476, 20, 2015.
98. Vijayakumar, M., Bhuvaneswari, M., Nachimuthu, P., Schwenzer, B., Kim, S., Yang, Z., Liu, J., Graff, G.L., Thevuthasan, S., Hu, J., Spectroscopic investigations of the fouling process on Nafion membranes in vanadium redox flow batteries. *J. Membr. Sci.*, 366, 325, 2011.
99. Vijayakumar, M., Bhuvaneswari, M., Nachimuthu, P., Schwenzer, B., Kim, S., Yang, Z., Liu, J., Graff, G.L., Thevuthasan, S., Hu, J., Spectroscopic investigations of the fouling process on Nafion membranes in vanadium redox flow batteries. *J. Membr. Sci.*, 366, 325, 2011.
100. Leung, P.K., Xu, Q., Zhao, T.S., Zeng, L., Zhang, C., Preparation of silica nanocomposite anion exchange membranes with low vanadium-ion crossover for vanadium redox flow batteries. *Electrochim. Acta*, 105, 584, 2013.
101. Teng, X., Sun, C., Dai, J., Liu, H., Su, J., Li, F., Solution casting Nafion/polytetrafluoroethylene membrane for vanadium redox flow battery application. *Electrochim. Acta*, 88 :725, 2013.
102. Mohammadi, T., Skyllas-Kazacos, M., Modification of anion-exchange membranes for vanadium redox flow battery applications. *J. Power Sources*, 63, 179, 1996.
103. Schwenzer, B., Zhang, J., Kim, S., Li, L., Liu, J., Yang, Z., Membrane development for vanadium redox flow batteries. *ChemSusChem.*, 4, 1388, 2011.
104. Won, S., Oh, K., Ju, H., Numerical analysis of vanadium crossover effects in all–vanadium redox flow batteries. *Electrochim. Acta*, 177, 310, 2015.
105. Ahn, S.M., Jeong, H.Y., Jang, J.K., Lee, J.Y., So, S., Kim, Y.J., Hong, Y.T., Kim, T.H., Poly- benzimidazole/Nafion hybrid membrane with improved chemical stability for vanadium redox flow battery application. *RSC Adv.*, 2018, 8, 25304, 2018.
106. Mai, Z., Zhang, H., Li, X., Xiao, S., Zhang, H., Nafion/polyvinylidene fluoride blend membranes with improved ion selectivity for vanadium redox flow battery application. *J. Power Sources*, 196, 5737, 2011.
107. Teng, X., Sun, C., Dai, J., Liu, H., Su, J., Li, F., Solution casting Nafion/polytetrafluoroethylene membrane for vanadium redox flow battery application. *Electrochim. Acta*, 88, 725, 2013.
108. Li, X., Zhang, H., Mai, Z., Zhang, H., Vankelecom, I. Ion exchange membranes for vanadium redox flow battery (VRB) applications. *Energy Environ. Sci.*, 4, 2011.
109. Prifti, H., Parasuraman, A., Winardi, S., Lim, T.M., Skyllas-Kazacos, M., Membranes for redox flow battery applications. *Membranes*, 2, 275 2012.
110. Yue, M., Zhang, Y., Wang, L., Sulfonated polyimide/chitosan composite mem- brane for vanadium redox flow battery: Membrane preparation, characterization, and single cell performance. *J. Appl. Polym. Sci.*, 127, 4150, 2012.

111. Chen, D., Hickner, M.A., Wang, S., Pan, J., Min, X., Meng, Y., Directly fluorinated polyaromatic composite membranes for vanadium redox flow batteries. *J. Membr. Sci.*, 139, 415, 2012.
112. Dai, J., Teng, X., Song, Y., Jiang, X., Yin, G., A super thin polytetrafluoroethylene/sulfonated poly(ether ether ketone) membrane with 91% energy efficiency and high stability for vanadium redox flow battery. *J. Appl. Polym. Sci.*, 133, 2016.
113. Yuan, Z., Dai, Q., Lin, Q., Zhao, Y., Zhang, H., Li, X., Highly stable aromatic poly (ether sulfone) composite ion exchange membrane for vanadium flow battery. *J. Membr. Sci.*, 541, 2017.
114. Qiu, J., Zhao, L., Zhai, M., Ni, J., Zhou, H., Peng, J., Li, J., Wei, G., Pre-irradiation grafting of styrene and maleic anhydride onto PVDF membrane and sub-sequent sulfonation for application in vanadium redox batteries. *J. Power Sources*, 177, 617, 2008.
115. Zhang, F., Zhang, H., Qu, C., A dication cross-linked composite anion-exchange membrane for all-vanadium flow battery applications. *ChemSusChem.*, 6, 2290, 2013.
116. Ahn, Y., Kim, D., Anion exchange membrane prepared from imidazolium grafted poly (arylene ether ketone) with enhanced durability for vanadium redox flow battery. *J. Ind. Eng. Chem.*, 71, 361, 2019.
117. Yuan, J., Yu, C., Peng, J., Wang, Y., Ma, J., Qiu, J., Li, J., Zhai, M., Facile synthesis of amphoteric ion exchange membrane by radiation grafting of sodium styrene sulfonate and N, N-dimethylaminoethyl methacrylate for vanadium redox flow battery. *J. Polym. Sci., Part A: Polym. Chem.*, 51, 5194, 2013.
118. Lu, W., Yuan, Z., Li, M., Li X., Zhang, H., Vankelecom, I., Solvent-induced rearrangement of ion-transport channels: A way to create advanced porous membranes for vanadium flow batteries. *Adv. Funct. Mater.*, 27, 2017.
119. Peng, S., Wu, X., Yan, X., Gao, L., Zhu, Y., Zhang, D., Li, J., Wang, Q., He, G. Poly-benzimidazole membranes with nanophase-separated structure induced by non-ionic hydrophilic side chains for vanadium flow batteries. *J. Mater. Chem. A*, 6, 3895, 2018.
120. Chen, D., Qi. H., Sun, T., Yan, C., He, Y., Kang, C., Yuan, Z., Li, X., Polybenzimidazole membrane with dual proton transport channels for vanadium flow battery application. *J. Membr. Sci.*, 586, 202, 2019.
121. Chae, I.S., Luo, T., Moon, G.H., Ogieglo, W., Kang, Y.S., Wessling, M., Ultra-high pro- ton/vanadium delectivity for hydrophobic polymer membranes with intrinsic nanopores for redox flow battery. *Adv. Energy Mater.*, 6, 2016.
122. Liu, L., Wang, C., He, Z., Das, R., Dong, B., Xie, X., & Guo, Z. An overview of amphoteric ion exchange membranes for vanadium redox flow batteries. *J. Mater. Sci. Technol.*, 69, 212-227, 2021.
123. Kong, L., Zheng, L., Niu, R., Wang, H., & Shi, H. A sulfonated poly (ether ether ketone)/amine-functionalized graphene oxide hybrid membrane for vanadium redox flow batteries. *RSC Adv.*, 6, 100262-100270, 2016.

124. Yang, X., Yan, L., Ran, F., Pal, A., Long, J., & Shao, L., Interface-confined surface engineering constructing water-unidirectional Janus membrane. *J. Membr. Sci.*, 576, 9-16, 2019.
125. Li, S., Jiang, X., Yang, X., Bai, Y., & Shao, L., Nanoporous framework "reservoir" maximizing low-molecular-weight enhancer impregnation into CO2-philic membranes for highly-efficient CO2 capture. *J. Membr. Sci.*, 570, 278-285, 2019.
126. Yang, X., Wang, Z., & Shao, L. Construction of oil-unidirectional membrane for integrated oil collection with lossless transportation and oil-in-water emulsion purification. *J. Membr. Sci.*, 549, 67-74, 2018.
127. Xu, T. Ion exchange membranes: State of their development and perspective. *J. Membr. Sci.*, 263, 1-29, 2005.
128. Ran, J., Wu, L., He, Y., Yang, Z., Wang, Y., Jiang, C., Xu, T., Ion exchange membranes: New developments and applications. *J. Membr. Sci.*, 522, 267-291, 2017.
129. Dai, J., Dong, Y., Yu, C., Liu, Y., & Teng, X. A novel Nafion-g-PSBMA membrane prepared by grafting zwitterionic SBMA onto Nafion via SI-ATRP for vanadium redox flow battery application. *J. Membr. Sci.*, 554, 324-330, 2018.
130. Lin, C. H., Yang, M. C., & Wei, H. J. Amino-silica modified Nafion membrane for vanadium redox flow battery. *J. Power Sources*, 282, 562-571, 2015.
131. Wang, G., Zhang, M., He, Z., Zhang, J., Chen, J., Wang, R., ... & Dai, Y. Novel amphoteric ion exchange membranes by blending sulfonated poly (ether ether ketone) with ammonium polyphosphate for vanadium redox flow battery applications. *J. Appl. Poly. Sci.*, 138, 50592, 2021.
132. Liu, Y., Dai, G., Chen, Y., Wang, R., Li, H., Shi, X., ... & Zhao, Y. Effective Design Strategy of Small Bipolar Molecules through Fused Conjugation toward 2.5 V Based Redox Flow Batteries. *ACS Energy Lett.*, 7, 1274-1283, 2022.
133. Qian, P., Wang, H., Zhang, L., Zhou, Y., & Shi, H., An enhanced stability and efficiency of SPEEK-based composite membrane influenced by amphoteric side-chain polymer for vanadium redox flow battery. *J. Membr. Sci.*, 643, 120011, 2022.
134. Sun, C., & Zhang, H., Review of the Development of First-Generation Redox Flow Batteries: Iron-Chromium System. *ChemSusChem*, 15, e202101798, 2022.
135. Zhou X.L., Zhao T.S., An L., Wei L., Zhang C., The use of polybenzimidazole membranes in vanadium redox flow batteries leading to increased coulombic efficiency and cycling performance, *Electrochim. Acta.*, 153, 492-498, 2015.
136. Aili D., Yang J., Jankova K., Henkensmeier D., Li Q., From polybenzimidazoles to polybenzimidazoliums and polybenzimidazolides, *J. Mater. Chem. A.*, 8, 12854-12886, 2020.
137. Choi S.-Y., Cho S., Kim D., Kim J., Song G., Singh R., Kim C., Boosting the proton conduction using protonated imidazole for advanced ion conducting membrane, *J. Membr. Sci.* 620, 118904, 2021.

138. Agmon N., The Grotthuss mechanism, *Chem. Phys. Lett.*, 244, 5-6, 456-462, 1995.
139. Yuan Z., Duan Y., Zhang H., Li X., Zhang H., Vankelecom I., Advanced Porous Membranes With Ultra-High Selectivity and Stability for Vanadium Flow Batteries, *Energy Environ. Sci.*, 9, 441-447, 2016.
140. Yuan Z., Duan Y., Liu T., Zhang H., Li X., Toward a low-cost alkaline zinc-iron flow battery with a polybenzimidazole custom membrane for stationary energy storage, *iScience*, 3, 40–49, 2018.
141. Kraglund M. R., Carmo M., Schiller G., Ansar S. A., Aili D., Christensen E., J. O. Jensen, Ion-solvating membranes as a new approach towards high rate alkaline electrolyzers, *Energy Environ. Sci.*, 12, 3313-3318, 2019.
142. Jensen J. O., Aili D., Hansen M. K., Li Q., Bjerrum N. J., Christensen E., A stability study of alkali doped PBI membranes for alkaline electrolyzer cells. *ECS Trans.*, 64, 1175, 2014.
143. Aili D., Wright A. G., Kraglund M. R., Jankova K., Holdcroft S., Jensen J. O., Towards a stable ion-solvating polymer electrolyte for advanced alkaline water electrolysis, *J. Mater. Chem. A*, 5, 5055-5066, 2017.
144. Peng, S., Yan, X., Zhang, D., Wu, X., Luo, Y., & He, G., A H3PO4 pre-swelling strategy to enhance the proton conductivity of a H 2 SO 4-doped polybenzimidazole membrane for vanadium flow batteries. *RSC Adv.*, 6, 23479-23488, 2016.
145. Bouchet, R.; Miller, S.; Duclot, M.; Souquet, J.L., A thermodynamic approach to proton conductivity in acid-doped polybenzimidazole. *Solid State Ion.* 145, 69-78, 2001.
146. Baozhong, X.; Savadogo, O., The effect of acid doping on the conductivity of polybenzimidazole (PBI). *J. New Mater. Electrochem. Syst.*, 2, 95-101, 1999.
147. Dai, Q., Xing, F., Liu, X., Shi, D., Deng, C., Zhao, Z., Li, X., High-performance PBI membranes for flow batteries: from the transport mechanism to the pilot plant. *Energy Environ. Sci.* 2022.
148. Wang, J.T.; Lin, W.F.; Weber, M.; Wasmus, S.; Savinell, R.F., Trimethoxymethane as an alternative fuel for a direct oxidation PBI polymer electrolyte fuel cell. *Electrochim. Acta.*, 43, 3821-3828, 1998.
149. Herranz, D.; Escudero-Cid, R.; Montiel, M.; Palacio, C.; Fatás, E.; Ocón, P., Poly (vinyl alcohol) and poly (benzimidazole) blend membranes for high performance alkaline direct ethanol fuel cells. *Renew. Energy*, 127, 883-895, 2018.
150. Glipa, X.; Bonnet, B.; Mula, B.; Jones, D.J.; Rozière, J., Investigation of the conduction properties of phosphoric and sulfuric acid doped polybenzimidazole. *J. Mater. Chem.*, 9, 3045-3049, 1999.
151. Ikhsan, M. M., Abbas, S., Do, X. H., Choi, S. Y., Azizi, K., Hjuler, H. A., Henkensmeier, D., Polybenzimidazole membranes for vanadium redox flow batteries: Effect of sulfuric acid doping conditions. *Chem. Eng. J.*, 435, 134902, 2022.
152. Ma, Y.L., Wainright, J.S., Litt, M.H., Savinell, R.F., Conductivity of PBI membranes for high-temperature polymer electrolyte fuel cells. *J. Electrochem. Soc.*, 151, 8-16, 2004.

153. Glipa, X., Bonnet, B., Mula, B., Jones, D.J., Rozière, J., Investigation of the conduction properties of phosphoric and sulfuric acid doped polybenzimidazole. *J. Mater. Chem.*, 9, 3045-3049, 1999.
154. Asensio, J.A.; Borrós, S.; Gómez-Romero, P., Polymer electrolyte fuel cells dased on phosphoric acid-impregnated poly (2,5- benzimidazole) membranes. *J. Electrochem. Soc.*, 151, 304-310, 2004.
155. Staiti, P., Lufrano, F., Aricò, A.S., Passalacqua, E., Antonucci, V., Sulfonated polybenzimidazole membranes—preparation and physico-chemical characterization. *J. Membr. Sci.*, 188, 71-78, 2001.
156. Bae, J.M., Honma, I., Murata, M., Yamamoto, T., Rikukawa, M., Ogata, N., Properties of selected sulfonated polymers as proton-conducting electrolytes for polymer electrolyte fuel cells. *Solid State Ion.*, 147, 189-194, 2002.
157. Asensio, J.A., Borrós, S., Gómez-Romero, P., Proton-conducting polymers based on benzimidazoles and sulfonated benzimidazoles. *J. Polym. Sci. Pol. Chem.*, 40, 3703–3710, 2002.
158. Wu, X., & Scott, K., A H2SO4 loaded polybenzimidazole (PBI) membrane for high temperature PEMFC. *Fuel Cells*, 12, 583-588, 2012.
159. Xu, H., Chen, K., Guo, X., Fang, J., Yin, J., Synthesis of novel sulfonated polybenzimidazole and preparation of cross-linked membranes for fuel cell application. *Polymer*, 48, 5556-5564, 2007.
160. Savadogo, O., Varela, F.J.R., Low temperature direct propane polymer electrolyte membranes fuel cell (DPFC). *J. New Mater. Electrochem. Syst.*, 4, 93-98, 2001.
161. Zhou, Z., Zholobko, O., Wu, X. F., Aulich, T., Thakare, J., & Hurley, J., Polybenzimidazole-Based Polymer Electrolyte Membranes for High-Temperature Fuel Cells: Current Status and Prospects. *Energies*, 14, 135, 2020.

2

Electrolytes Materials for Redox Flow Batteries

Mandira Mitra, Sandip Kundu, Mousumi Layek, Subhodip Mondal, Ujjwal Mandal* and Bidyut Saha†

Department of Chemistry, The University of Burdwan, Burdwan, West Bengal, India

Abstract

The rising demand to accumulate growing amounts of sustainably produced electrical energy has sparked a surge in the exploration of redox flow batteries (RFBs). When paired with photovoltaic and wind farms, RFBs become attractive prospects for the establishment of smart grids. In recent years, the investigation has turned from materials made of metal to active materials consisting of organic substances to fulfil the objective of "green," safe, and cost-effective energy storage. Commercial RFBs employ low-cost redox-active inorganic ions with high ionic activity and a commercial ion-exchange medium, but they have several flaws, such as expensive and hazardous active materials, redox species crossover, and high ion exchange membrane expenses. Since minimal polymeric redox-active substances and size-exclusion membranes with porosity produced from generic polymers are available which might be used to substitute exclusive inorganic active ingredients and membranes having the capability of exchanging ions, the inclusion of active polymeric substituents offers an appealing approach. This chapter presents an overview of the various redox flow batteries, their historical development, and battery performance. Their advantages and limitations are also discussed.

Keywords: Redox, catholyte, anolyte, battery, storage

**Corresponding author*: umandal@chem.buruniv.ac.in (U. Mandal)
†*Corresponding author*: bsaha@chem.buruniv.ac.in (B. Saha)

Inamuddin and Tariq Altalhi (eds.) Advanced Redox Flow Technology, (51–82) © 2024 Scrivener Publishing LLC

2.1 Introduction

Nowadays the extreme exhaustion of non-renewable sources of energy and deliverance of superfluous CO_2 in the environment due to the combustion of fossil fuels make it necessary to focus the worldwide science community on thinking about renewable sources of energy. Generally, research gradually shifted towards various renewable sources of energy like a nuclear power source. Again, a safety issue arises for renewable nuclear power, bearing in mind the accident in the nuclear power plant in Japan in 2011. Such incidents drew the attention of scientists toward harmless, cleaner, and reproducible energy resources. Some renewable sources are sunlight, wind, tidal energy, geothermal energy, etc. However, another problem arises about renewable sources regarding their discontinuous and capricious nature in terms of availability, which restricts their efficacy as a trustworthy power supply. Solar energy, for example, is not 100% reliable and there is a high initial capital cost for the plant. In the case of wind energy, the noise effect generated due to the wind turbines is a matter of concern. Wind energy is squandered due to the incapability to ship wind power during the night when the generation of wind is at its peak but consumer response is the lowest; consequently, the addition of storage adds substantial value. Most renewable sources are unpredictable. For that, only a small percentage of the energy generated from primarily renewable sources can be utilized by mankind. It has been discovered that if the renewables which are not dispatchable generate more than 20% of total energy deprived of being connected to energy storage, the electrical grid may be disrupted [1, 2]. Again, photovoltaic and solar-energy deployment will necessitate arbitrage because, while solar irradiation obtained terrestrially in around a single hour is adequate to supply global energy needs for a year provided that the sun does set every day. Accordingly, the need to deduce renewable energy resources with greater potential and to assimilate them proficiently into the electric grid in a huge capacity generates an optimistic field in the research of the energy storage system. Presently the Pump Hydroelectric Energy Storage (PHES) dominates at the ground with a huge amount of power-loading capacity but it has some severe drawback that limits its proficient use, such as high capital costs for the installation and PHES is largely circumscribed by geographical conditions. Redox Flow Batteries (RFB) are the utmost prominent, versatile, and profitable option for extensive storage of energy, and they are an appealing technology because of their independent control over energy and power, flexible design, cheap cost, and safety features [3]. Because the quantity of stored energy is determined by the reactant species employed, and the reactant tank size, and concentration,

RFBs propose the key benefit that the outputs of power and energy are autonomous variables [4-6].

2.2 Overview of Redox Flow Battery

RFB is an electrochemical energy storage device and the core component is the electrochemical cell. In the electrochemical cell, RFB converts the power into the fluids having lively redox character and accumulates that into the exterior tank present on the outer side of the electrochemical cell like fuel cells. This is the main reason behind the separate grid storage application of RFBs. In a usual RFB cell, for the redox reaction to occur, graphite felt, metal foam, or carbon paper electrode surfaces are obligatory. Redox-active constituents are solubilized into the positive electrolyte namely catholyte and negative electrolyte namely anolyte solution (see Figure 2.1). While charging catholyte materials get oxidized by losing electrons, anolyte materials become reduced by accepting electrons [7]. The peristaltic pump fitted with the setup circulates the discharged electrolytes in the reservoirs and the electrolytes which are charged die down to reservoirs. The electrochemical processes are reversible during the discharge phase. To establish charge balance, subsidiary ions present in electrolyte solutions infiltrate with the help of the separator during the procedure. A separator performs the following three functions: (a) prevents redox-active molecules from crossing between the catholyte and anolyte sides; (b) conserves charge balance by allowing diffusion of supporting ions; (c) prevents a cell's internal short circuit. Electrolyte reservoir volume, and redox-active species concentration within, besides reduction potential differences of the species which store energy all balance with this design, although the power measures predominantly with the dynamic area in the electrochemical cell load [8]. Unlike an enclosed battery, this allows energy and power to be adjusted freely. As a result, for the high ratio of energy to power (i.e., extended release periods following ranked capacity), the arrangement budget depends not on the electrochemical cells but on the electrolyte itself [9]. One of the most challenging technical issues in RFBs is preventing electro-active materials from crossing over. Such crossover of the electroactive materials decreases the Coulombic Efficiency (CE) as well as worsens the capacity of the battery. To nullify such a serious issue and to improve the performance of the membrane separator two approaches generally have been adopted. The first option is to reduce the membrane's pore size or increase the molecular size of electroactive components to increase the membrane's physical obstructive property [10-12]. Yu and co-workers

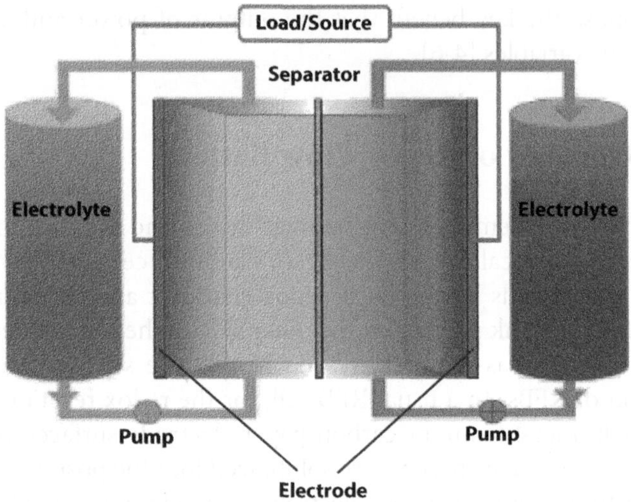

Figure 2.1 Redox Flow Battery (schematic representation) [23].

proposed a membrane based on graphene oxide and an organometallic structure-based spongy membrane for RFBs, which includes the creation of a nanochannel structure that prevents electroactive species crossover and increases charge carrier permeability [12, 13]. Donnan exclusion is enhanced in the second route [14]. The Donnan exclusion effect is a term used to explain how charged ions are distributed along an organism's permeable membrane. This principle is also known as the Gibbs–Donnan effect [15]. The ionic exchange membrane can prevent charged electroactive molecules from crossing across. This phenomenon is effective such that an anion exchange membrane containing a positively charged environment can suppress the passage of the catholyte particles by electrostatic repulsion and similarly the cation exchange membrane having a negative charge can prevent the passage of the anolyte particles.

2.3 Measurement of the Capacity of the Redox Flow Battery

Several characteristics, including, energy density, power density, and others, are necessary for the measurement of the functioning of flow batteries. For RFBs power density is calculated by measuring the power of the RFB per unit contracting area of the electrode with the separator. This can be represented by the following equation:

$$\text{Power density} = \frac{V \times I}{S} \quad \text{(eq. 2.1)}$$

Here V is the voltage of the battery in the volt unit used and I represent the current in the ampere unit and S is the contracting area in the m^2 unit.

However, power density for solid-state batteries is commonly represented as the weight of the solid battery per volume of the electrolyte. Due to polarization, the discharge voltage of an RFB often differs from the potential theoretically. The overpotential is equally important for reaction kinetics, redox species diffusion coefficient in electrolytes, and separator ionic conductivity along with the discharge current density. The reaction rate of three factors—the redox substances, separator ionic conductivity, and battery voltage—generally contribute to determining the power density.

The flow battery's energy density is another essential component in determining its efficiency. It determines the capacity of energy per unit mass (Ec/W) or per unit volume (Ec/L). In RFBs the energy is deposited in the liquid state electrolytes, hence per unit volume energy is pragmatic. The following equation is used to compute energy density:

$$\text{Energy density} = FU\frac{nCV}{V_{+} + V_{-}} \quad \text{(eq. 2.2)}$$

Here, U represents the battery's voltage (Volt) and n represents electrons engaged during the redox process, C denotes concentration (Molarity), and F represents Faraday's constant. The volumes of catholyte and anolyte are represented by V_+ and V_-, respectively. Therefore the battery voltage will be the compactness of energy inside the cell and the potential difference between anolyte and catholyte is proportional to voltage. The total energy concreteness with numerous potentials will be equal to the sum of the individual potentials' energy density of the battery [16].

Battery efficiency can be determined by CE, Voltage Efficiency (VE), and Energy Efficiency (EE).

$$\text{CE} = \frac{Q_{discharge}}{Q_{charge}} \quad \text{(eq. 2.3)}$$

$$\text{EE} = \frac{E_{discharge}}{E_{charge}} \quad \text{(eq. 2.4)}$$

Here Q means the quantity of electric charge and E denotes energy. EE to CE ratio denotes VE.

$$VE = \frac{EE}{CE} \tag{eq. 2.5}$$

Generally, the RFBs do not attain 100% efficiency due to some of their major drawbacks, such as at different charged states the redox species become unstable, the crossover of the anolyte and catholyte and ionic conductivity of the electrolyte, etc. Along with these factors, the lifetime of the battery also plays an important role in determining the stability of the RFBs.

Furthermore, the battery's life span is significant for any kind of energy storage device. The battery's life span is determined by the electroactive element's stability in all charged states.

2.4 Formation of Redox-Active Constituents for RFB

RFB research is mainly reliant upon the creation of active redox substituents. RFB was initially proposed in the 19th century, presenting inorganic species like chlorides of vanadium or titanium [17]. In 1970 the National Aeronautics and Space Administration (NASA) focused on the development of redox-active materials like iron (Fe^{2+}/Fe^{3+}), titanium (Ti^{3+}/TiO^{+}), and chromium (Cr^{2+}/Cr^{3+}) systems. After such efforts, in 1980 Skyllas-Kazacos and co-workers shaped the all-vanadium redox flow battery (VRFB) which turned into the most studied and prosperous RFB technology so far [18-20]. Most significantly, the catholyte and anolyte materials show an important character in RFBs. We can classify the redox flow batteries based on the electrolytes used in the cell into inorganic, organic, and polymeric RFBs. Reaction kinetics of the redox-active materials largely depends on the appropriate density of current, and the electrolyte's viscosity has also been hampered by the nature and the concentration of the redox-active material. At higher viscosity higher energy is required for the circulation of the charges and charge carrier mobility of the electrolytes decreased. Generally, the active redox metals are the metal-based redox couple that is solubilized in the aqueous medium but a large number of energy-storing materials which generally depend on organic redox-active compounds have recently been identified [21, 22]. The utilization of the organic solvent is sometimes necessary as not all of the organic redox active materials dissolve in the water. Organic aprotic solvents have a larger potential window and higher electrochemical stability than protic solvents like water [23]. Stable nitrosyl (NO) radicals, carbonyl (CO) compounds, and organosulphur compounds, for example, all have been explored as potential energy storage components in organic batteries based on polymer [24]. These organic active materials can be obtained synthetically and should feature two stable chemically reversibly redox states. Organic redox active materials can be synthesized in properly equipped laboratories. For that, countries having a lack of raw materials can satisfy their requirement for these organic materials from their chemical industry.

2.4.1 Inorganic Redox Flow Battery

Traditional RFBs generally employ redox-active inorganic materials in aqueous solutions. Inorganic RFBs are categorized into iron/chromium, all-vanadium, polysulfide/bromine, vanadium/bromine, etc. Some other RFBs also have been reported such as titanium/manganese, zinc/cerium, vanadium/cerium, single metal RFBs, etc. Among these types, some of the RFBs are discussed in this section.

2.4.1.1 All Vanadium RFBs

An all-vanadium system is the most advanced of all types of RFBs; it has passed the demonstration stage and now is ready for commercialization [25]. Because vanadium exists in a variety of oxidation states, it may function simultaneously as an anolyte and catholyte in both of the half cells. Redox Flow Battery made using Vanadium (VRFB) was invented mainly to resolve difficulties related to cross-contamination produced over various ionic substances diffusing through the membrane. Cross-contamination would still occur theoretically because both half cells contain almost the same chemical, but only as a result of self-discharge, allowing the electrolyte life to be endlessly extended [25]. Redox couples including vanadium were useful for the application of redox cells by NASA, but they were shelved because of the very low soluble nature of vanadium pentoxide (V_2O_5). For that vanadium, electrolyte concentration become lower than 0.5 M which was too small to utilize practically [26]. Skyllas-Kazacos and colleagues electrochemically oxidize the vanadyl sulfate containing V^{4+} ion which is highly soluble to the V^{5+} ion in the sulphuric acid solution having the strength of 2(M), therefore solving a problem related to solubility and thus improving the efficiency of the VRFB [27-30]. Furthermore, instability, elevated cost, and the toxic nature of the compounds consisting of vanadium, as well as the deployment of acidic electrolytes with high concentrations, prevent VRFBs from being widely used [7].

In the anode and cathode half-cells, VRFB uses the V^{2+}/V^{3+} and V^{4+}/V^{5+} couples with the following charge-discharge reactions:

$$\text{Cathode reaction: } VO_2^+ + 2H^+ + e^- \rightleftharpoons VO^{2+} + H_2O \qquad E^0 = 1.00\ V\ vs\ SHE \quad \text{(eq. 2.6)}$$

$$\rightleftharpoons\text{Anode reaction: } V^{2+} - e- \rightleftharpoons V^{3+} \quad E^0 = -0.25\ V\ vs\ SHE \quad \text{(eq. 2.7)}$$

$$\text{Overall reaction: } VO_2^+ + V^{2+} + 2H^+ \rightleftharpoons VO^{2+} + V^{3+} + H_2O \quad E^0 = 1.25\ V\ vs\ SHE \quad \text{(eq. 2.8)}$$

2.4.1.2 *Zinc/Bromine RFBs*

The powerful catalyst [25-29], highly active electrode [30-37], advanced modeling [38-43], and innovative flow field design [44-47] of the redox flow batteries using vanadium have been extensively examined in recent decades, but commercialization has been hampered due to the expensive nature of the vanadium materials and a narrow operating temperature window. Because of its higher energy density (70 $Whkg^{-1}$) and cheaper cost compared to the other alternatives, the Zinc-Bromine Redox Flow Battery (ZBRFB) is ideally suited for extensive stationary uses as an energy storage technology. Although the Zn/Br system uses inexpensive active elements, the low conductivity and slow kinetics of the neutral electrolyte result in poor battery performance [48]. In the case of ZBRFB zinc metal is used as the anolyte material; this is appealing because zinc metal has a significant volumetric capacity [49]. As Zn/Zn^{2+} combination has a modest redox potential (0.76 V), ZBRFB has a theoretic potential value of 1.84 V. Typical reactions occurring in electrodes of a ZBRFB are as follows:

$$\text{Cathode reaction: } Br_2 + 2e^- \rightleftharpoons 2Br^- \quad E^0 = 1.08\ V \quad \text{(eq. 2.9)}$$

$$\text{Anode reaction: } Zn - 2e^- \rightleftharpoons Zn^{2+} \quad E^0 = -0.76\ V \quad \text{(eq. 2.10)}$$

Overall reaction: $Br_2 + Zn \rightleftharpoons 2Br^- + Zn^{2+}$ $E^0 = 1.84\ V$ (eq. 2.11)

2.4.1.3 Tin/Bromine Redox Flow Battery

There have already been proposals for flow batteries involving Sn/V and Sn/Fe redox couple [50, 51]. However, the crossover of Sn, V, and Fe ions caused capability degradation throughout the battery operation. Previously mixed solutions are commonly used to avoid cross-contamination, but they quadruple the cost of redox materials and diminish redox couple solubility in the electrolyte. Yikai and colleagues introduce a new redox flow battery consisting of tin and bromine, i.e., Sn/Br RFB that uses the negative, as well as positive redox couples Sn/Sn2+ and Br/Br2 correspondingly, and the Sn/Br RFB, was put through a 500-cycle test at 120 $mAcm^{-2}$ and 35°C, as illustrated in Figure 2.2a and Figure 2.2b, which show the reverse electro-deposition process of Sn for battery capacity recovery [52]. To report the issue of cross-contamination, a reverse-electrodeposition approach using Sn is developed and as illustrated in Figure 2.2c, a mechanism was developed to transport intersected Sn^{2+} starting with the positive counterpart to negative along with attaining inside the recovery of the ability of the cell during the function of the battery. During the cycle test, the CE, VE, and EE remain constant. However, as illustrated in Figure 2.2d, Sn ions on the negative side of the membrane migrate to the positive side which causes capacity degradation of the redox flow battery.

The working principle of Br/Sn RFB is as follows:

Cathode reaction: $Br_2 + 2e^- \rightleftharpoons 2Br^-$ $E^0 = 1.09\ V\ vs\ SHE$ (eq. 2.12)

Anode reaction: $Sn + 2e^- \rightleftharpoons Sn^{2+}$ $E^0 - -0.14\ V\ vs\ SHE$ (eq. 2.13)

Overall reaction: $Sn + Br_2 \rightleftharpoons Sn^{2+} + 2Br^-$ $E^0 = 1.23\ V\ vs\ SHE$ (eq. 2.14)

While charging, on the positive side the Br^- ions are oxidized by losing an electron to form Br_3^- or Br_2 and the Sn^{2+} ions on the negative side receive electrons from the positive electrode via the exterior circuit and deposit as metallic Sn on the negative electrode and the process is reversed during the discharging process.

The Sn/Br RFBs' charge-discharge polarisation has been investigated using various previously treated electrodes at low current densities and

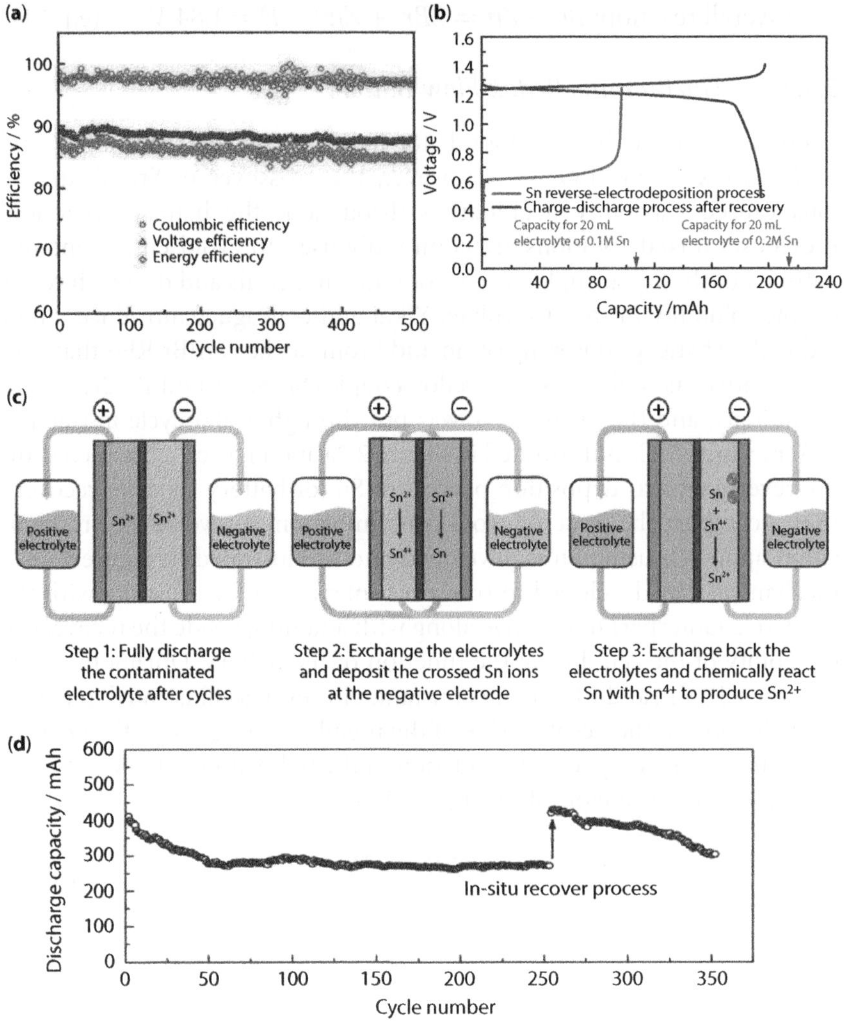

Figure 2.2 Sn/Br RFB's cycle performance at 120 mAcm−2 and 35°C; (b) Reverse-electrodeposition curve for battery capacity recovery of Sn; (c) For battery capacity recovery process steps and (d) battery capacity retained in the *in-situ* recovery process during cycle test [52].

overpotentials by Zeng *et al.* [52]. At 50% state of change and 25°C, Figure 2.3a displays the IR-free polarisation battery curves along with KOH pre-treatment, warm air pre-treated, and untouched electrodes. In this case, the battery's concentration loss may be neglected since there is a sufficient supply of the active species at a small value of operating current densities. As a result, the IR-free polarizations are essentially identical to the activation loss [52].

The charging and discharging curves are shown in Figure 2.3b. As shown in Figure 2.3c, the VE touches 92.4 percent at 80 mAcm^{-2} and stays at 84.7% at an extraordinary current density of 200 mAcm^{-2}. From 80 to 200 mA cm^{-2}, the CE remains over 95%. At 200 and 80 mAcm^{-2}, the Sn/Br RFB has an EE of 82.7% and 87.9%, respectively. Furthermore, when the current density is 80 mA cm^{-2}, consumption of the electrolyte stretches 92.0%, but reduces to 59.6% at 200 mA cm^{-2}, necessitating additional inquiry into flow rate optimization and cell structure modification for mass transfer

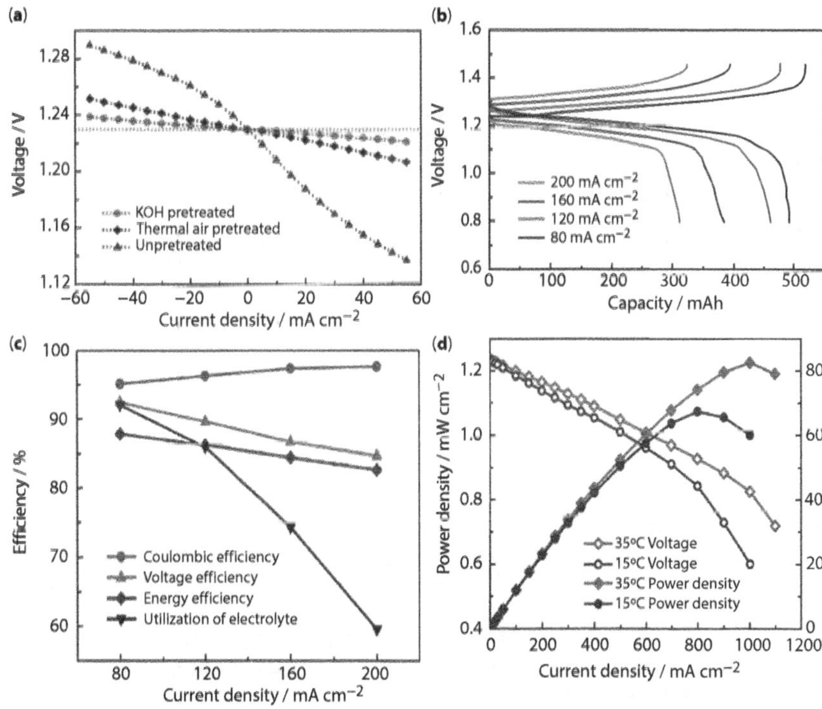

Figure 2.3 (a) At low current densities, IR-free polarisation curves of Sn/Br RFBs with various pre-treatment electrodes at normal temperature; (b) charging-discharging plots and (c) using KOH pre-treated electrode efficacies of the Sn/Br RFB at the temperature 35 °C; and (d) Sn/Br RFB complete discharging polarisation curves [52].

augmentation. Figure 2.3d shows the battery's full discharge polarisation characterization at 15°C and 35°C with a 50% state of change.

In addition, as a closed system, the flow battery, the Sn/Br electrolyte may be recycled indefinitely. In an acidic solution, both Sn/Sn^{2+} and $Br^-/$ Br2 redox couples are largely soluble and have simple kinetics [51, 53]. As a result, electrolytes' impact on the environment is mostly unseen.

2.4.1.4 *Iron-Chromium RFB*

Redox Flow Battery consisting of Iron and Chromium (ICRFB) is the most initially synthesized RFB and it utilizes the soluble redox couples of Fe^{3+}/Fe^{2+} and Cr^{3+}/Cr^{2+} in the acid-supporting medium as positive and negative

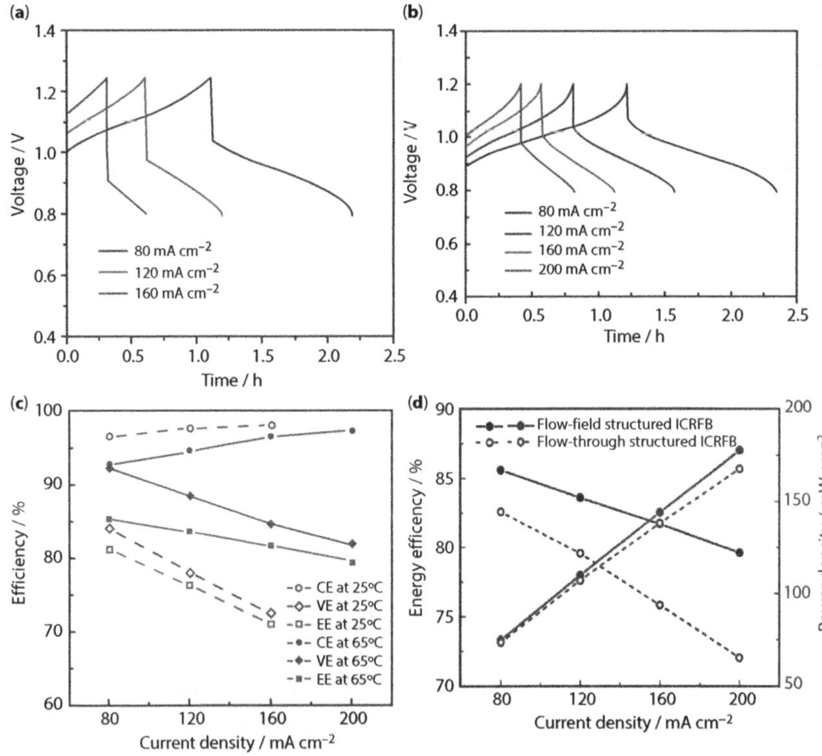

Figure 2.4 Charging-discharging curves of the iron/chromium redox flow battery at (a) 25oC and (b) 65oC; (c) efficiencies of the iron/chromium redox flow battery; and (d) At 65oC, energy efficiency and middling discharging power thicknesses for the current and conservative iron/chromium redox flow batteries [54].

electrolytes [54]. The storing and releasing of the electrical energy by ICRFB occur through the following reactions:

$$\text{Cathode reaction: } Fe^{3+} + e^{-} \rightleftharpoons Fe^{2+} \quad E^0 = 0.77\ V\ vs\ SHE \quad \text{(eq. 2.15)}$$

$$\text{Anode reaction: } Cr^{2+} - e- \rightleftharpoons Cr^{3+} \quad E^0 = -0.41\ V\ vs\ SHE \quad \text{(eq. 2.16)}$$

$$\text{Overall reaction: } Fe^{3+} + Cr^{2}+ \rightleftharpoons Fe^{2+} + Cr^{3+} \quad E^0 = 1.18\ V\ vs\ SHE \quad \text{(eq. 2.17)}$$

Figure 2.4 shows the analysis of testing the flow-field organized ICRFB at various current densities changing with the temperatures.

The ICRFB has traditionally used a straightforward flow-through cell construction, in which the electrolyte is pushed directly through the porous electrode. Zeng and co-workers found that the typical ICRFB's high ohmic resistance severely restricts the performance of the battery [54, 55]. ICRFB can boost the battery performance in which a thinner electrode is installed. To disperse the electrolyte throughout the whole electrode region, the flow-field compartment arrangement is constructed using distinct flow networks [56]. In comparison to the iron redox couple on the carbonaceous electrode, the chromium redox couple possesses weak reversibility and sluggish kinetics. Furthermore, during the charging process of the cell evolution of hydrogen on the Cr^{3+}/Cr^{2+} side, a suitable catalyst is required, which is often expensive [57, 58]. Another problem is that the absence of an effective practitioner with good selectivity for ions causes considerable crossover of the electrolytes. The aging impact of active cross-species transport in a cell containing a couple of half cells, like in the Fe/Cr redox flow system, is caused by a degradation of the number of vigorous substances and therefore degrades the capability of the cell [59, 60].

2.4.1.5 Polysulfide-Bromine RFB

The anolyte and catholyte in a Polysulfide Bromine Redox Flow Battery (PSRFB) are a sulfide-polysulfide couple and a bromine couple, respectively [61]. A cation exchange membrane separates the two electrodes. In lithium-sulfur batteries, the elemental sulfur (S_8) has been considered as a cathode [67, 68]. The principal source of deterioration in a lithium-sulfur battery is the "shuttle" effect of low oxidation state sulfur (polysulphides, S_x^{2-}; 1< x <8). Considering sulfide ion as for illustration, the persulphide-sulfide redox couple represents a −0.51 V potential, which has a theoretic potential of 1.59 V.

$$\text{Cathode reaction: } Br_2 + 2e^- \rightleftharpoons 2Br^- \quad E^0 = 1.08\ V\ vs\ SHE \quad \text{(eq. 2.18)}$$

$$\text{Anode reaction: } 2S^{2-} - 2e^- \rightleftharpoons S_2^{2-} \quad E^0 = -0.51\ V\ vs\ SHE \quad \text{(eq. 2.19)}$$

$$\text{Overall reaction: } Br_2 + 2S^{2-} \rightleftharpoons 2Br^- + S_2^{2-} \quad E^0 = 1.59\ V\ vs\ SHE \quad \text{(eq. 2.20)}$$

PSB, like other redox flow batteries, has several benefits over traditional rechargeable batteries, including the ability to independently scale up power range, intense discharge capabilities, and extended shelf life [62-66]. Furthermore, the chemical reagents employed in PSRFBs' electrolytes are abundant and inexpensive as well as easily dissolved in an aqueous solution and reasonably benign, making PSRFB much more cost-effective for amplifying energy-storing capability. The PSRFB system, however, struggles from the crossover of the solutions containing electrolytes via the membrane because of the employment of two distinct solutions in each half-cell.

2.4.1.6 *Titanium-Manganese Redox Flow Battery*

Among various possibilities, a low-cost possibility is a titanium-manganese electrolyte. Titanium and manganese are both inexpensive materials that are abundant in nature. The electromotive force in a sulphuric acid solution is 1.41 V, which is high among aqueous electrolytes, as shown by the cell reaction equations [69].

$$\text{Cathode reaction: } Ti^{3+} + H_2O \rightleftharpoons TiO^{2+} + 2H^+ + e^- \quad E^0 = 0.1\ V\ vs\ SHE \quad \text{(eq. 2.21)}$$

$$\text{Anode reaction: } Mn^{3+} + e^- \rightleftharpoons Mn^{2+} \quad E^0 = 1.51\ V\ vs\ SHE \quad \text{(eq. 2.22)}$$

$$\text{Overall reaction: } Ti^{3+} + Mn^{3+} + H_2O \rightleftharpoons TiO^{2+} + Mn^{2+} + 2H^+ E^0 = 1.41\ V\ vs\ SHE \quad \text{(eq. 2.23)}$$

The precipitation of manganese dioxides (MnO_2) happens, in reality, owing to the disproportionation process during charging, because the Mn^{3+} ion is chemically unstable in an aqueous solution which can be understood by the following reaction:

$$2Mn^{3+} + 2H_2O \rightleftharpoons Mn^{2+} + MnO_2 + 4H^+$$

Dong *et al.* used the working electrode and counter electrode both made of carbon felt, and the electrolyte at the working electrode was maintained static while the electrolyte at the counter electrode flowed during the measurement [69]. The battery capacity may be diminished and the cell stack may become clogged if manganese dioxides collect in the tank. When using this material in redox flow batteries, several precautions must be taken to keep the Mn^{3+} ion stable and to prevent the precipitation of MnO_2. From the above equation of the disproportionation reaction, it is clear that a highly acidic medium and high concentration of Mn^{2+} will favor the stability of Mn^{3+}. However, even using these approaches, stabilizing Mn^{3+} ions at large concentrations is challenging, and as a result, the battery's energy density is restricted. Now it has been discovered that Titanium-Manganese electrolyte can suppress the precipitation of MnO_2 [70, 71].

2.4.2 Organic Redox Flow Battery

Rather than metal ion active materials, organic materials have been used in the development of redox flow batteries in recent years. The advantages of utilizing organic redox-active materials are due to their variety at the molecular level, structural design ability, and cheap cost. When we discuss "all-organic" or "organic" redox flow batteries, here the discussion is not on the supporting electrolyte or the solvent but about the redox active materials which could be an aqueous solvent or some organic solvent having the capability to solubilize active organic redox materials. Organic aprotic solvents have a larger potential window and higher electrochemical stability than protic solvents like water. Because higher-voltage redox pairs might be used, this could lead to higher-energy-density batteries. Organic solvents, on the other hand, have a far lower ion conductivity, limiting the current densities that may be used. A greater voltage somewhat mitigates this impact on power density. Some organic compounds, especially those with polar substituents, are soluble in water. As a result, huge current densities may be used, but the voltage between the two redox partners is limited. As a result, an "L" shape curve can be seen in Figure 2.5, illustrating the limits of present flow-battery technology [23].

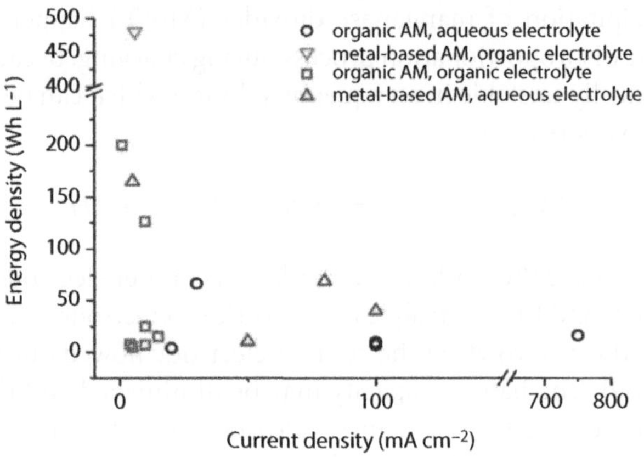

Figure 2.5 Density of energy vs. density of current plot of selected redox flow battery systems (AM=active material) [23].

Organic reagents such as the compounds which can form complexes [72, 73] were used at the initial stage in the previous decade; after that, metal-ligand complexes are used as the energy storage materials in which the organic ligands are used as the energy storage materials [74, 75], and finally, flow batteries having an organic and an inorganic active material [76, 77].

Stable nitrosyl radicals, carbonyl compounds, and organosulphur compounds have been explored as potential energy storage components in organic batteries based on polymer which can be utilized as a starting point for developing organic materials that are acceptable for use in RFBs [24]. These materials must have somewhat two stable and electrochemically reversible redox states, which might be generated using synthetic methods. Petrochemistry can supply the starting materials and some other sources like extraction from renewable sources and white biotechnology can be used for the generation of starting materials [75]. Nonetheless, using the organic beginning constituents is not confined to a definite geographic location, but may be synthesized everywhere provided autonomous organic raw material acquisition is available. Despite a shortage of certain basic raw components, some countries may profit from their harsh chemical manufacturing units, which may manufacture active organic compounds in a huge proportion. An additional benefit of organic materials is their adaptability. To increase battery performance, the organic core component can be modified with certain functional groups to promote chemical reversibility, solubility, and redox potential.

2.4.2.1 Quinone-Based Redox Active Materials

Xu and co-workers constructed the first organic/inorganic flow battery in 2009 [78]. The anode-active material in the single-flow cadmium/chloranil flow battery was metallic cadmium, while the anolyte was an acidified $CdSO_4$ solution. Here organic cathode was tetrachloro-p-benzoquinone (chloranil, QCl_4) which is insoluble.

$$\text{Cathode reaction: } QCl_4 + 2H^+ + 2e^- \rightleftharpoons H_2QCl_4 \quad E^0 = 0\ .71\ V \tag{eq. 2.24}$$

$$\text{Anode reaction: } Cd \rightleftharpoons Cd^{2+} + 2e^- \quad E^0 = -0.40\ V \tag{eq. 2.25}$$

$$\text{Overall reaction: } QCl_4 + 2H^+ + Cd \rightleftharpoons H_2QCl_4 + Cd^{2+} \quad E^0 = 1.11\ V \tag{eq. 2.26}$$

The claimed single-flow battery takes the benefit of being free from the membrane, which saves money in the long run. On the other hand, chloranil cathode's demand for very acidic electrolytes might be considered a disadvantage.

Xu *et al.* stated a Tiron (lead/4, 5-dibenzoquinone-1,3-benzenesulfonate) Hybrid Flow Battery (HFB) in a subsequent investigation which was fitted with a Nafion 115 membrane, a particular concentration of iron dissolved in H_2SO_4 employed for instance cathode stuff and the anode stuff is made of the lead couple [78]. A schematic representation of the HFB is shown in Figure 2.6. Tiron's electrochemical behavior is significantly pH-dependent, similar to chloranil and it is reported that only electrolytes with a pH of less than 4 can be used in the HFB. Furthermore, during the initial electrolysis, iron showed an electrochemical hydroxylation, which triggers an irreversible side reaction in a portion of the active material.

In 2012, Wang and colleagues presented an HFB consisting of both organic and inorganic materials built by a product derived from anthraquinone with methoxytriethyleneglycol molecules which are solubilized in the propylene carbonate and serve as the catholyte constituent organic in nature and lithium serve as the negative electrode. In the following arrangement LiPF2 act as the supporting electrolyte [79]. The redox mechanism of anthraquinone derivatives has been widely studied since they are renowned for using organic cathode materials utilized in lithium batteries

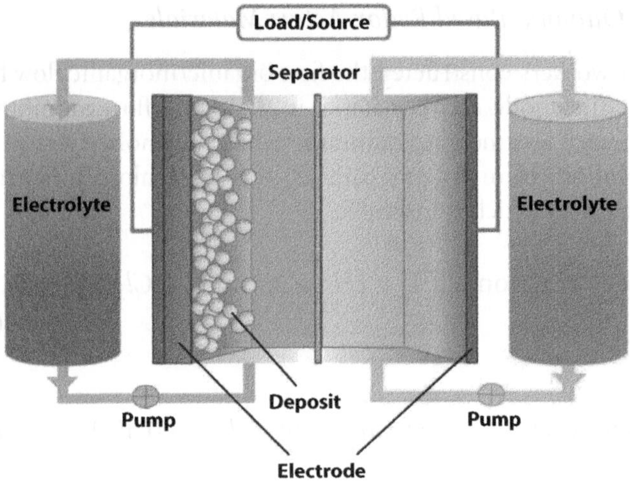

Figure 2.6 Graphic representation of a hybrid flow battery [23].

[79-84]. The ethylene glycol moieties help with solubility, especially in non-aqueous polar solvents, and a concentration of 0.25m has been achieved [75].

By using Density Functional Theory (DFT) calculations, Bachman and co-workers studied the redox chemical behavior, the impact of substances on their potential value, and Gibbs's free energy value for the solvation of roughly fifty types of derivatives of anthraquinone. They also showed that entirely methylating the core of anthraquinone declines potential value and the substituents which have a comparable redox range will enhance solvation energy [85].

2.4.2.2 Tempo-Based Redox-Active Materials

TEMPO (2,2,6,6-tetramethylpiperidine-1-oxyl) is a red-orange solid that may be sublimed under vacuum; its melting point is 36°C-38°C and it is a stable nitroxide radical with heterocyclic moiety. It has been investigated as an active redox material for both lithium generally semi-organic and all-organic radical containing batteries [86-88]. Wei *et al.* published a new non-aqueous electrolyte in 2014 that contained high concentrations of TEMPO as the electroactive material and a large extent hexafluorophosphate of Lithium (LiPF6) which act as the secondary charge shipper into solvent consisting of propylene carbonate (PC), ethylene carbonate (EC), and ethyl methyl carbonate (EMC) in the proportion of 1:4:5 [76]. Due to the stable, quick, and reversible nature of non-aqueous electrolytes,

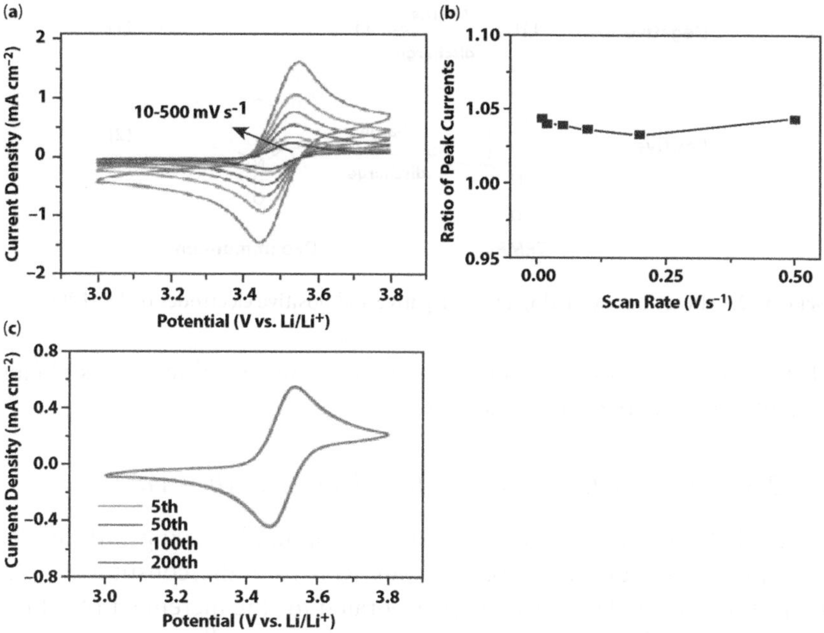

Figure 2.7 (a) Cyclic Voltametry curves of TEMPO in LiPF6 solution; (b) in terms of scan rate, the fraction of oxidation and reduction peak currents; (c) Cyclic Voltammetry curves at 50 mVs–1 [26].

the polymeric cathode materials based on TEMPO have high rate ability, a proper redox potential, and strong long-term cycle stability [89-91].

Cyclic Voltametries were implemented in the solution having TEMPO (0.005 M) and $LiPF_6$ (1.0 M) dissolved in PC, EC, and EMC for the identification of the redox pairs and assess the reversible nature of the redox reactions (1:4:5 by weight). At scan speeds ranging from 10 to 500 mVs^{-1}, TEMPO exhibits definite peaks (Figure 2.7a). In scheme 2.1 the reactions occurring in the negative and positive electrodes are shown and the transfer of one electron takes place there. The ratio of the peak current in the cathode and anode compartment is extremely near to one at entirely scan speeds, as shown in Figure 2.7b, suggesting good reversibility of the redox process. At a scan rate of 50 mVs^{-1}, the CV curves of the 5th, 50th, 100th, and 200th cycles almost fully overlap (Figure 2.7c), confirming the TEMPO redox couple's great stability [76]. TEMPO is a good choice for flow batteries due to its great reversible and stable nature of the redox process.

The organic RFB flow cell system containing Li uses the organic TEMPO catholyte in combination with a Li-metal anode, which overtakes other

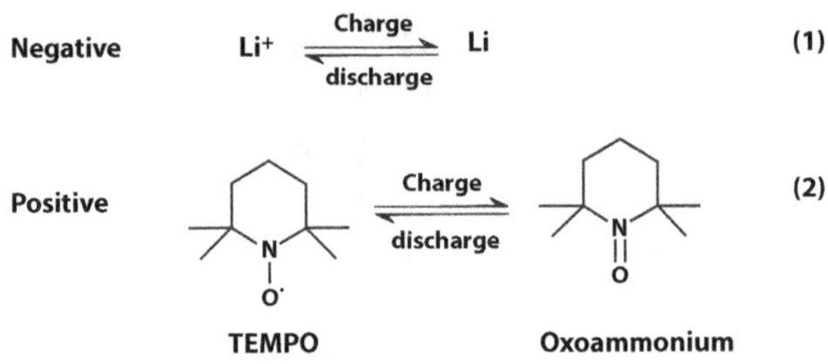

Scheme 2.1 Reactions occurring in the negative and positive electrodes of TEMPO.

RFBs in terms of both concentration and voltage, resulting in a sufficient potential volumetric energy density.

2.4.2.3 *Redox Active Materials Based on Alkoxybenzene*

In 2014, Huang *et al.* specified active compounds having little molecular weight, for example non-aqueous RFB cathodic constituents [92]. Oligoethylene oxide chains are incorporated for the increment of solubility and energy density into the dimethoxy-di-tertbutylbenzene which is in charge of various electrochemical characteristics and the steadiness of that material. Aside from improved solubility, the scientists noticed a change in physical characteristics at normal temperature, moving from a solid state compound 2,5-ditertbutyl-1,4-bis(2-methoxy-ethoxy)benzene (DBBB) to a semi-liquid (ANL-10) to a liquid state (ANL-8, ANL-9), the structures are shown in the Figure 2.8.

When cyclic voltammetry is applied over the propylene carbonate using $LiBF_4$, the three energy depository materials (ANL-10, ANL-9, and

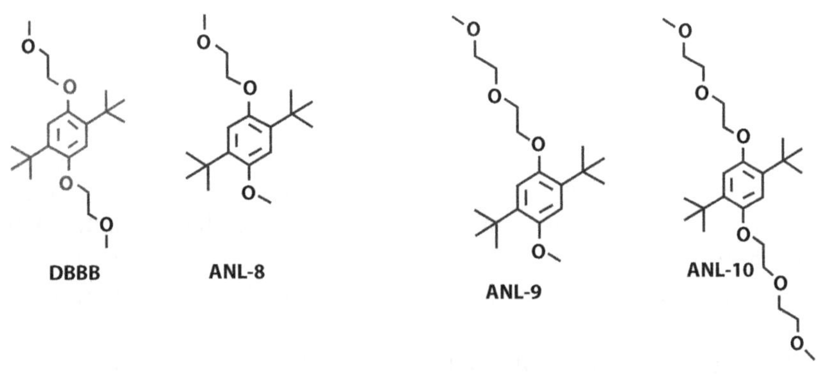

Figure 2.8 Structures of 2, 5-di-tert-butyl-1, 4-bis (2-methoxy-ethoxy) benzene (DBBB) and reformed products ANL-8, ANL-9, and ANL-10.

ANL-8) showed strong reversible electrochemical character and potentials of roughly 4.0 V when conjugated with Li^+ /Li. Moreover, ANL-8 has a diffusion coefficient on an equal level with vanadium [19, 93], DBBB [94], and numerous quinone derivatives [95]. The battery experiments indicated various issues about the chemical reversibility of the active material, despite the alkoxybenzenes strong electrochemical characteristics as evidenced in CV measurements. As a result, their applicability as active materials is debatable.

2.5 Hybrid Electrolytes Used in a Lithium Redox Flow Battery

In 2011 Goodenough's and Zhou's teams independently proposed the new perception of building a flow battery scheme using metallic Lithium (or sometimes metallic Sodium) as an anode, in addition to a flow-through water solution as a cathode [96-99]. The electrochemical reactions that occur at the cathode and anode are straightforward and may be summarised as follows:

$$\text{Cathode reaction: } M_{(aq)}^{z+} + ne^- \rightleftharpoons M_{(aq)}^{(z-n)+}$$

Where M denotes a metal element, z represents the valance of the metal ion.

$$\text{Anode reaction: } nLi \rightleftharpoons nLi^+ + ne^-$$

$$\text{Overall reaction: } M_{(aq)}^{z+} + nLi \rightleftharpoons M_{(aq)}^{(z-n)+} + nLi^+$$

The basic setup of the two organizations' prototype battery systems is substantially the same. In the anode section, metallic Li is inserted as an anode in an electrolyte organic in nature. There is a stirring aqueous solution at the cathode chamber that functions as both a cathode and a catholyte at the same time. Between the cathode and anode, a thin plate called LISICON functioned as a separating plate. The battery includes the benefits of an advanced self-regulating process on energy capacity and power from a RFB, by merging metallic Li with an aqueous cathode circulating in nature [1].

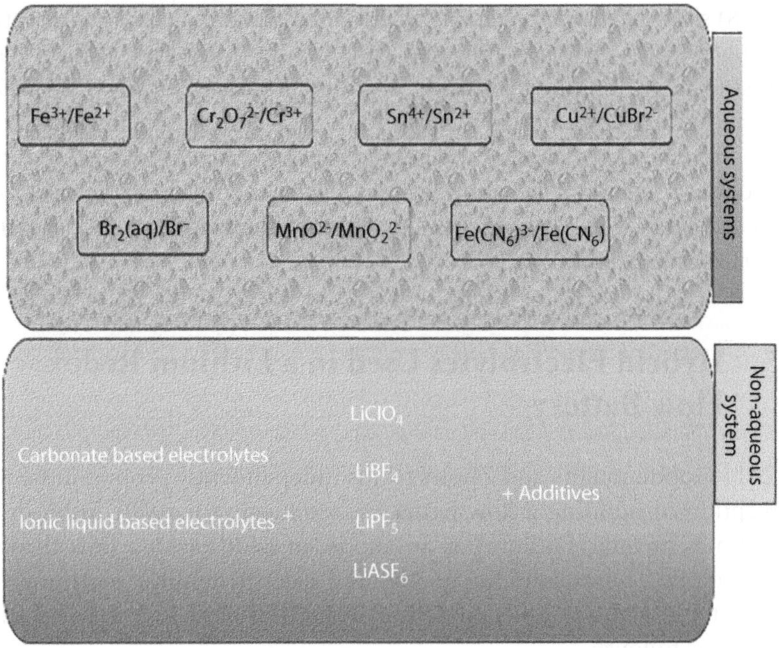

Figure 2.9 Lithium–water battery systems. (The standard electrode potentials of the redox couplings are shown numerically in the image.). Reprinted with consent [97].

A flow battery system consisting of Lithium metal relies on an electrolyte having solid nature generally termed LISICON (Lithium Super Ionic Conductor) plate which separates organic compounds on the anode from

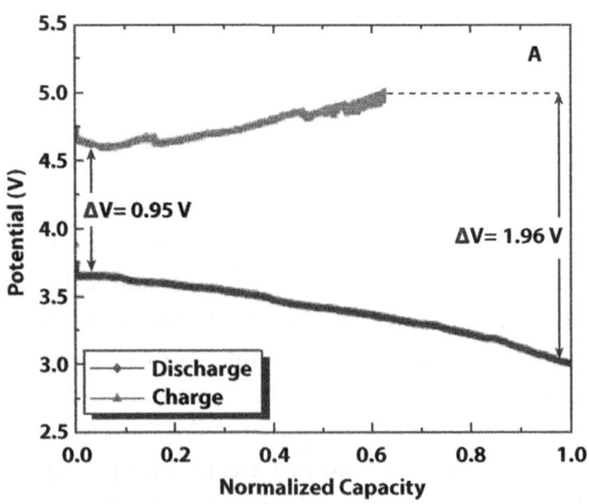

Figure 2.10 Charge/discharge curves for the Li/Fe3+ (aq) battery in 2M Fe (NO3)3+ & 1M LiNO3 cathode solution and 0.5 mA current [96].

the cathode side where an aqueous mixture is present. The solid electrolyte must be steady across the pH range.

The choice of the aqueous cathode is also important. The aqua electrodes incorporated in redox flow batteries of Li are essential to have the subsequent properties, as stated by Lu & Goodenough *et al.* [94-96]: (1) correct electrode potential values; (2) no adjacent reactions; (3) stable enough in an aqueous medium; (4) reversible; (5) consistent protection; and (6) inexpensive price. Also, they have presented a list of potential aqueous cathode redox partners [97]. (See Figure 2.9).

If the lifetime of a commercial solid electrolyte is investigated as well as the water decomposition process, it's understandable that the aqueous cathode's voltage versus Li redox couple will be limited narrowing redox couples' options. An aqueous nitrate solution of Fe^{3+}/Fe^{2+} was used by Goodenough *et al.* as the liquid cathode in their original proof of concept investigation [94, 96]. They measured discharging voltages and charging voltages versus Li redox couple (Figure 2.10). A comparatively wide difference between the charging and discharging curves indicates that the battery is inefficient. Also, they have claimed that the battery consists of a short phase life, and according to them this is due to the acidic nature of the fluid, which attacked the solid electrolyte Ti (IV). In dilute aqueous solution, Fe^{3+} is recognized as vulnerable to sluggish hydrolysis, resulting in the solution's acidic character. When Fe^{3+} undergoes reduction to form Fe^{2+}, the acidity of the solution decreases, causing the precipitation of ferric hydroxide, this phenomenon lowers the active compound in the solution and therefore lowers the capability, as well as covers the superficial area of the LISICON plate, therefore increasing resistance. Consequently, in this scenario, the alkaline medium is insignificant.

2.6 Levelised Cost of the Redox Active Materials

A reduction in costs is an essential, if not the most important, enabling factor for RFB market penetration. For the charge storage procedure, organic compounds or earth-abundant metals can be used to achieve this purpose. Additionally, the use of simple separators rather than complicated ion-selective membranes, as well as the use of water as a solvent, would cut expenditures. Even though redox-active vanadium species are indefinitely stable, vanadium expenses are too volatile and, on average, too expensive for widespread use in large-scale RFB systems. RFB chemistry is based on more Earth-abundant basic elements in the case of the inorganic materials resulting in the development of the following substances such as hydrogen-bromine [100-101], metal-hydride [102], sulfur-oxygen [103], and slurry-based all-iron [103-105] topologies. Regardless of the little expected chemical expenses of these systems, they all face significant technological problems that prohibit them from replacing vanadium RFBs at this time. Non-aqueous RFBs can achieve much greater voltages and energy densities, but they use low-conductivity, volatile, and costly solvents.

The utilization of organic and organometallic redox active reactants is an alternate route to low-cost RFBs. But only some of the organic redox-active materials are commercially accessible; others must be manufactured using time-consuming and multi-step processes, making them even more expensive than standard inorganic redox-active materials. Scientists might be able to provide some insight into the most cost-effective synthesis method. Another problem is that capacity fade rates produced by the breakdown of organic elements are a major flaw in most organic chemistries that have been described; however, diffusion of dynamic substances across the membrane present also shows a significant character in various investigations. For setting up RFB with expected project lives of eras, such larger capability rates fading are simply unfeasible. However, for economically viable aqueous organic RFB arrangements, a decadal chemical lifetime is not required. Furthermore, given the potential of regularly refilling the lost electrolyte, a more acceptable objective is to ensure adequate constancy at a fair balanced charge of the loading chemical.

2.7 Conclusion

Flow batteries have several advantages. By scaling the tank volume and the cell loads separately, the system's power and capacity may be scaled independently for the exact adaptation of the linked generated unit. Flow batteries are primarily designed to store electricity for several hours and can change from charging to discharging in a fraction of a second [106]. One of the primary properties of RFBs that indicates substantial potential for stationary applications is the lack of physical material transfer through the electrode/electrolyte boundary in many configurations. While certain designs can be categorized as flow batteries simply because the active constituent runs from the cell's external to the surface of the electrode, the majority of redox flow battery systems in advance collect chemical energy. The electrochemical reactant species are solubilized in solutions and held outside cisterns, rather than being maintained inside the electrode, as intercalation electrodes or metal/metal alloy are present there. Because the reduced and versions oxidised of each reactant have good solubility in an electrolyte solution, they may be transported in the same phase to and from the electrode surface. During charge and discharge, simply the comparative amount of oxidized and reduced forms varies at every flow. Because the energy amount deposited is determined by the reactants utilized, their size, and the concentration of the tanks containing reactants, RFBs propose the distinct benefit that energy and power outputs are independent factors [4, 6].

The use of redox coupled with high reversibility having no parallel reactions and proposing many more charge/discharge cycles can increase life span. As a result, systems that simply transfer an electron without changing their shape are likely to be more stable. Redox reactions or pH-sensitive compounds should be avoided because the pH-sensitive groups can convert into another group like the ester group can undergo hydrolysis. In summary, research and development should emphasize organic active materials which are utilized by hybrid-flow batteries and aqueous redox flow batteries. Organic chemicals having high solubility, which can be manufactured with little effort and in a large number of yields, are preferred in this situation to allow for efficient manufacturing systems which are used in a great quantity with capabilities in MWh array. In addition, specific emphasis on the safety of the system, particularly in terms of toxicity and flammability, is necessary. Safe battery solutions are especially important for RFBs which could be used in households and may be utilized in conjunction with decentralized roof-top solar power production.

References

1. Wang, Y., He, P., Zhou. H., A review on Li-redox flow batteries based on hybrid electrolytes: At the cross road between Li-ion and redox flow batteries. *Adv. Energy Mater.*, 2, 770, 2012.
2. Weber, A.Z., Mench, M.M., Meyers, J.P., Ross, P.N., Gostick, J.T., Liu, Q.H., A review on redox flow batteries *J. Appl. Electrochem.* 41, 1137, 2011.
3. Lai, Y.Y., Lai, X., Zhu, Y., A review on polymeric active materials for redox flow battery application. *ACS Appl. Polym.* 2, 113, 2020.
4. Skyllas-Kazacos, M., Grossmith, F., Efficient vanadium redox flow cell. *J. Electrochem. Soc.*, 134, 2950, 1987.
5. Remick, R.J., Ang, P.G.P., Electrically rechargeable anionically active reduction-oxidation electrical storage-supply system. US Patent 5963537, assigned to National Power PLC, 1984.
6. Zhou, H., Zhang, H.M., Zhao, P., Yi, B.L., A comparative study of carbon felt and activated carbon based electrodes for sodium polysulfide/bromine redox flow battery *Electrochim. Acta* 51, 6304, 2006.
7. Chai, J., Lashgari, A., Jiang, J.J., Electroactive materials for next-generation redox flow batteries: from inorganic to organic, pp. 1-47, ACS, 2020.
8. Kwabi, D.G., Ji, Y., Aziz, M.J., Electrolyte lifetime in aqueous organic redox flow batteries: a critical review. *Chem. Rev.* 120, 646, 2020

9. Li, Z., Sam, P.M., Su, L., Tsai, P., Badel, A.F., Valle, J.M., Eiler, S.L., Xiang, K., Brushett, F.R., Chiang, Y., Air-breathing aqueous sulfur flow battery for ultralow-cost long-duration electrical storage, *Joule.*, 1, 306, 2017.
10. Shrestha, A., Hendriks, K., Sigman, M., Minteer, S., Sanford, M., Realization of an asymmetric non-aqueous redox flow battery through molecular design to minimize active species crossover and decomposition. *Chem. Eur. J.*, 26, 5369, 2020.
11. Zhang, L., Ding, Y., Zhang,C., Zhou, Y., Zhou, X., Liu, Z., Yu, G., Enabling graphene-oxidebased membranes for large-scale energy storage by controlling hydrophilic microstructures. *Chem.*, 4, 1035, 2018.
12. Nagarjuna, G., Hui, J., Cheng, K. J., Lichtenstein, T., Shen, M., Moore, J. S., Rodríguez López, J., Impact of redox-active polymer molecular weight on the electrochemical properties and transport across porous separators in non-aqueous solvents. *J. Am. Chem. Soc.* 136, 16309, 2014.
13. Peng, S., Zhang, L., Zhang, C., Ding, Y., Guo, X., He, G., Yu, G., Gradient-distributed metal–organic framework–based porous membranes for non-aqueous redox flow batteries. *Adv. Energy Mater.*, 8, 1802533, 2018.
14. Beh, E. S., De Porcellinis, D., Gracia, R. L., Xia, K. T., Gordon, R. G., Aziz, M. J., A neutral ph aqueous organic-organometallic redox flow battery with extremely high capacity retention. *ACS Energy Lett.*, 2, 639, 2017.
15. Zeng, L., Zhao, T.S., Wei, L., Jiang, H.R., Wu, M.C., Anion exchange membranes for aqueous acid-based redox flow batteries: current status and challenges. *Appl. Energy.*, 233, 622, 2019.
16. Potash, R. A., McKone, J. R., Conte, S., Abruña, H. D., On the benefits of a symmetric redox flow battery, *J. Electrochem. Soc.*, 163, A338, 2016.
17. Bartolozzi, M., Development of redox flow batteries. a historical bibliography. *J. Power Sources*, 27, 219, 1984.
18. Rychcik, M., Skyllas-Kazacos, M., Characteristics of a new all-vanadium redox flow battery, *J. Power Sources*, 22, 59,1988.
19. Sum, E., Rychcik, M., Skyllas-kazacos, M., Investigation of the V (V)/V (IV) system for use in the positive half-cell of a redox battery. *J. Power Sources*, 16, 85, 1985.
20. Sum, E., Skyllas-Kazacos, M., A study of the V (II)/V(III) redox couple for redox flow cell applications. *J. Power Sources*, 15, 179, 1985.
21. Janoschka,T., Martin, N., Hager, M.D., Schubert, U.S., An aqueous redox-flow battery with high capacity and power: The TEMPTMA/MV system. *Angew. Chem. Int. Ed.*, 55, 14427, 2016.
22. Winsberg, J., Stolze, C., Muench, S., Liedl, F., Hager, M.D., Schubert, U.S., *In-situ* polymerized binder: A three-in-one design strategy for all-integrated sio_x anode with high mass loading in lithium ion batteries. *ACS Energy Lett.*, 6, 290, 2016.
23. Winsberg, J., Hagemann, T., Janoschka, T., Hager, M.D., Schubert, U.S., Redox-flow batteries: From metals to organic redox active materials. *Angew. Chem. Int. Ed.*, 56, 686, 2011.

24. Muench, S., Wild, A., Friebe, C., Haeupler, B., Janoschka, T., Schubert, U.S., Polymer-based organic batteries. *Chem. Rev.*, 116, 9438, 2016.
25. Jiang, H., Shyy, W., Wu, M., Ruihan Z., Zhao T.S., A bi-porous graphite felt electrode with enhanced surface area and catalytic activity for vanadium redox flow batteries. *Appl. Energy*, 233,105, 2019.
26. Jiang, H., Wei, S., Lin, Z., Zhang, R., Zhao, T.S., Highly efficient and ultra-stable boron doped graphite felt electrodes for vanadium redox flow batteries. *J. Mater. Chem.*, 6, 12344, 2018.
27. Zeng, L., Zhao, T., Wei, L., Zeng, Y., Zhou, X., Mn_3O_4 nanoparticle-decorated carbon cloths with superior catalytic activity for the VII/VIII redox reaction in vanadium redox flow batteries. *Energy Technol.*, 6, 1228, 2018.
28. Wei, L., Zhao, T.S., Zhao, G., An, L., Zeng, L., A high-performance carbon nanoparticle decorated graphite felt electrode for vanadium redox flow batteries. *Appl. Energy*, 176, 74, 2016.
29. Wei, L., Zhao, T.S., Zeng, L., Zhou, X.L., Zeng, Y.K., Titanium carbide nanoparticle-decorated electrode enables significant enhancement in performance of all-vanadium redox flow batteries. *Energy Technol.*, 4, 990, 2016.
30. Sun, J., Zeng, L., Jiang, H., Chao, C., Zhao, T., Formation of electrodes by self-assembling porous carbon fibers into bundles for vanadium redox flow batteries. *J. Power Sources* 106, 405, 2018.
31. Wang, R., Li, Y., He, Y.L., Achieving gradient-pore-oriented graphite felt for vanadium redox flow batteries: meeting improved electrochemical activity and enhanced mass transport from nano-to micro-scale. *J. Mater. Chem. A*, 7, 10962, 2019.
32. Wu, Q., Zhang, X., Lv, Y., Lin, L., Liu, Y., Zhou, X., Bio-inspired multi-scale-pore-network structured carbon felt with enhanced mass transfer and activity for vanadium redox flow batteries. *J. Mater. Chem. A.*, 6, 20347, 2018.
33. Wang, Q., Qu, Z., Jiang, Z., Yang, W., Experimental study on the performance of a vanadium redox flow battery with non-uniformly compressed carbon felt electrode. *Appl. Energy*, 213, 293, 2018.
34. Jiang, H.R., Sun, J., Wei, L., Wu, M.C., Shyy, W., Zhao, T.S., A high power density and long cycle life vanadium redox flow battery. *Energy Storage Mater.*, 24, 529, 2020.
35. Wang, R., Li, Y., Twin-cocoon-derived self-standing nitrogen-oxygen-rich monolithic carbon material as the cost-effective electrode for redox flow batteries. *J. Power Sources*, 421, 139, 2019.
36. Wu, Q., Lv, Y., Lin, L., Zhang, X., Liu, Y., Zhou, X., An improved thin-film electrode for vanadium redox flow batteries enabled by a dual layered structure. *J. Power Sources*, 410, 152, 2019.
37. Zhou, X., Zhang, X., Lv, Y., Lin, L., Wu, Q., Nano-catalytic layer engraved carbon felt via copper oxide etching for vanadium redox flow batteries. *Carbon*, 153, 674, 2019.
38. Yang, W.W., He, Y.L., Li, Y.S., Performance modeling of a vanadium redox flow battery during discharging. *Electrochim. Acta*, 155, 279, 2015.

39. Wei, Z., Lim, T.M., Skyllas-Kazacos, M., Wai, N., Tseng, K.J., Online state of charge and model parameter co-estimation based on a novel multi-timescale estimator for vanadium redox flow battery. *Appl. Energy*, 172, 169, 2016.
40. Ke, X., Prahl, J.M., Alexander, J.I.D., Savinell, R.F., Mathematical modeling of electrolyte flow in a segment of flow channel over porous electrode layered system in vanadium flow battery with flow field design. *Electrochim. Acta*, 223, 124, 2017.
41. Chen, H., Li, X., Gao, H., Liu, J., Yan, C., Tang, A., Numerical modelling and in-depth analysis of multi-stack vanadium flow battery module incorporating transport delay. *Appl. Energy*, 247, 13, 2019.
42. Chen, H., Wang, S., Gao, H., Feng, X., Yan, C., Tang, A., Analysis and optimization of module layout for multi-stack vanadium flow battery module. *J. Power Sources*, 427, 154, 2019.
43. Yin, C., Guo, S., Fang, H., Liu, J., Li, Y., Tang, H., Numerical and experimental studies of stack shunt current for vanadium redox flow battery. *Appl. Energy*, 151, 237, 2015.
44. You, X., Ye, Q, Cheng P. Scale-up of high power density redox flow batteries by introducing inter digitated flow fields. *Int. Commun. Heat Mass Transf.*, 75, 7, 2016.
45. Zeng, Y., Li, F., Lu, F., Zhou, X., Yuan, Y., Cao, X., A hierarchical interdigitated flow field design for scale-up of high-performance redox flow batteries. *Appl. Energy*, 238, 435, 2019.
46. Sun, J., Zheng, M., Yang, Z., Yu, Z., Flow field design pathways from lab-scale toward large-scale flow batteries. *Energy*, 173, 637, 2019.
47. Yin. C., Gao. Y., Guo. S., Tang. H., A coupled three dimensional model of vanadium redox flow battery for flow field designs. *Energy*, 74, 886, 2014.
48. Wu, M., Zhao, T., Zhang, R., Jiang, H., Wei, L., A zinc-bromine flow battery with improved design of cell structure and electrodes. *Energy Technol.*, 6, 333, 2018.
49. Clark, S., Mainar, A. R., Iruin, E., Colmenares, L. C., Blázquez, J. A., Tolchard, J. R., Jusys, Z., Horstmann, B., Designing aqueous organic electrolytes for zinc–air batteries: method, simulation, and validation. *Adv. Energy Mater.* 10, 1903470, 2020.
50. Chen, F., Sun, Q., Gao, W., Liu, J., Yan, C., Liu, Q., Study on a high current density redox flow battery with tin (II)/tin as negative couple. *J. Power Sources*, 280, 227, 2015.
51. Zhou, X., Lin, L., Lv, Y., Zhang, X., Wu, Q., A Sn-Fe flow battery with excellent rate and cycle performance. *J. Power Sources*, 404, 89, 2018.
52. Zeng, Y., Yang, Z., Lu, F., Xie, Y., A novel tin-bromine redox flow battery for large-scale energy storage. *Appl. Energy*, 255, 113756, 2019.
53. Huskinson, B., Marshak, M.P., Suh, C., Er, S., Gerhardt, M.R., Galvin, C.J., Chen, X., Aspuru-Guzik, A., Gordon R.G., Aziz, M.J., A metal-free organic-inorganic aqueous flow battery. *Nature*, 505, 195, 2014.

54. Zeng, Y.K., Zhou, X.L., An, L., Wei, L., Zhao, T.S., A high-performance flow-field structured iron-chromium redox flow battery. *J. Power Sources*, 324, 738, 2016.
55. Wang, W., Luo, Q., Li, B., Wei, X., Li, L., Yang, Z., Recent progress in redox flow battery research and development. *Adv. Funct. Mater.*, 23, 970, 2013.
56. Darling, R.M., Perry, M.L., The influence of electrode and channel configurations on flow battery performance. *J. Electrochem. Soc.*, 161, A1381, 2014.
57. Lopezatalaya, M., Codina, G., Perez, J.R., Vazquez, J.L., Aldaz, A., Climent, M.A., Behaviour of the Cr(III)/Cr(II) reaction on gold-graphite electrodes. application to redox flow storage cell. *J. Power Sources*, 35, 225, 1991.
58. Lopezatalaya, M., Codina, G., J. R. Perez, J.R., Vazquez, J.Z., Aldaz, A., Optimization studies on a Fe/Cr redox flow battery. *J. Power Sources*, 39, 147, 1992.
59. Vafiadis, H., Skyllas-Kazacos, M., Evaluation of membranes for the novel vanadium bromine redox flow cell. *J. Membr. Sci.*, 279, 394, 2006.
60. Schulte, D., Drillkens, J., Schulte,B., Sauer, D.U., Nafion hybrid membranes for use in redox flow batteries. *J. Electrochem. Soc.*, 157280, A989, 2010.
61. Zhao, P., Zhang, H., Zhou, H., Yi, B., Nickel foam and carbon felt applications for sodium polysulfide/bromine redox flow battery electrodes. *Electrochim. Acta*, 51, 1091, 2005.
62. L.H. Thaller, Electrically rechargeable REDOX flow cell, US Patent 3996064, assigned to NASA, 1976.
63. Skyllas-Kazacos, M., Kasherman, D., Hong, D.R., Kazacos, M., Characteristics and performance of 1 kW UNSW vanadium redox battery. *J. Power Sources*, 35, 399, 1991.
64. Fabjan, C., Garche, J., Harrer, B., Jorissen, L., Kolbeck, C., Philippi, F., Tomazic, G., Wagner, F., The vanadium redox-battery: An efficient storage unit for photovoltaic systems. *Electrochim. Acta*, 47, 825, 2001.
65. Fang, B., Iwasa, S., Wei, Y., Arai, T., Kumagai, M., A study of the Ce(III)/Ce(IV) redox couple for redox flow battery application. *Electrochim. Acta*, 47, 3971, 2002.
66. Bae, C.H., Roberts, E.P.L., Dryfe, R.A.W., Chromium redox couples for application to redox flow batteries. *Electrochim. Acta*, 48, 279, 2002.
67. Bai, S., Liu, X., Zhu, K., Wu, S., Zhou, H., Metal-organic framework-based separator for lithium-sulfur batteries. *Nat. Energy*, 1, 16094, 2016.
68. Li, G., Lu, F., Dou, X., Wang, X., Luo,D., Sun, H., Yu, A., Chen, Z., Polysulfide regulation by the zwitterionic barrier toward durable lithium-sulfur batteries. *J. Am. Chem. Soc.*, 142, 3583, 2020.
69. Dong, Y., Kaku, H., Miyawaki, H., Tatsumi, R., Moriuchi, K., Shigematsu, T., Titanium-manganese electrolyte for redox flow battery. *SEI Tech. Rev.*, 84, 27, 2017.
70. Dong, Y., Kaku, H., Miyawaki, H., Tatsumi, R., Moriuchi, K., Shigematsua, T., Novel titanium/manganese redox flow battery. pp. 59-67, *ECS Trans.*, 2015.

71. Kaku, H., Dong, Y.R., Hanafusa, K., Moriuchi, K., Shigematsu, T., Effect of Ti(IV) Ion on Mn(III) Stability in Ti/Mn Electrolyte for Redox Flow Battery, pp. 1-9, ECS Trans., 2016.
72. Lai, Q., Zhang, H., Li, X., Zhang, L., Cheng, Y., A novel single flow zinc-bromine battery with improved energy density. *J. Power Sources*, 235, 1, 2013.
73. Wen, Y.H., Zhang, H.M., Qian, P., Zhou, H.T., Zhao, P., Yi, B.L., Yang, Y.S., Studies on iron (Fe^{3+}/Fe^{2+})-complex/bromine (Br_2/Br^-) redox flow cell in sodium acetate solution. *J. Electrochem. Soc.*, 153, A929, 2006.
74. Kim, J.-H., Kim, K.J., Park, M.-S., Lee, N.J., Hwang, U., Kim, H., Kim, Y.-J., Development of metal-based electrodes for non-aqueous redox flow batteries. *Electrochem. Commun.*, 13, 997, 2011.
75. Chakrabarti, M.H., Dryfe, R.A.W., Roberts, E.P.L., Evaluation of electrolytes for redox flow battery applications. *Electrochim. Acta*, 52, 2189, 2007.
76. Wei, X., Xu, W., Vijayakumar, M., Cosimbescu, L., Liu, T., Sprenkle, V., Wang, W., TEMPO-based catholyte for high-energy density non-aqueous redox flow batteries. *Adv. Mater.*, 26, 7649, 2014.
77. Winsberg, J., Janoschka, T., Morgenstern, S., Muench, S., Hagemann, T., Hauffman, G., Gohy, J.-F., Hager, M.D., Schubert, U.S., Poly (TEMPO)/zinc hybrid-flow battery: A novel, "green," high voltage, and safe energy storage system. *Adv. Mater.*, 28, 2238, 2016.
78. Xu, Y., Wen, Y.-H., Cheng, J., Cao, G.-P., Yang, Y.-S., A study of tiron in aqueous solutions for redox flow battery application. *Electrochim. Acta*, 55, 715, 2010.
79. Wang, W., Xu, W., Cosimbescu, L., Choi, D., Li, L., Yang, Z., Anthraquinone with tailored structure for a nonaqueous metal–organic redox flow battery. *Chem. Commun.*, 48, 6669, 2012.
80. Alt, H., Binder, H., Kçhling, A., Sandstede, G., Investigation into the use of quinone compounds-for battery cathodes. *Electrochim. Acta*, 17, 873, 1972.
81. Gall, T.L., Reiman, K.H., Grossel, M.C., Owen, J.R., Poly (2, 5-dihydroxy-1, 4-benzoquinone-3,6-methylene): a new organic polymer as positive electrode material for rechargeable lithium batteries. *J. Power Sources*, 119, 316, 2003.
82. Z. Song, H. Zhan, Y. Zhou, Anthraquinone based polymer as high performance cathode material for rechargeable lithium batteries *Chem. Commun.*, 448, 2009.
83. Lei, Z., Wei-kun, W., An-bang, W., Zhong-bao, Y., Shi, C., Yusheng, Y., A MC/AQ parasitic composite as cathode material for lithium battery. *J. Electrochem. Soc.*, 158, A991, 2011.
84. Xu, W., Read, A., Koech, P.K., Hu, D., Wang, C., Xiao, J., Padmaperuma, A.B., Graff, G.L., Liu, J., Zhang, J.-G., Factors affecting the battery performance of anthraquinone-based organic cathode materials. *J. Mater. Chem.*, 22, 4032, 2012.
85. Bachman, J.E., Curtiss, L.A., Assary, R.S., Investigation of the redox chemistry of anthraquinone derivatives using density functional theory. *J. Phys. Chem. A*, 118, 8852, 2014.

86. Nishide, H., Iwasa, S., Pu, Y.-J., Suga, T., Nakahara, K., Satoh, M., Organic radical battery: Nitroxide polymers as a cathode-active material. *Electrochim. Acta*, 50, 827, 2004.
87. Liang, Y., Tao, Z., Chen, J., Organic electrode materials for rechargeable lithium batteries. *Adv. Energy Mater.*, 2, 742, 2012.
88. Janoschka, T., Hager, M.D., Schubert, U.S., Powering up the future: Radical polymers for battery applications. *Adv. Mater.*, 24, 6397, 2012.
89. Huang, Q., Choi, D.W., Cosimbescu, L., Lemmon, J.P., Multi-electron redox reaction of an organic radical cathode induced by a mesopore carbon network with nitroxide polymers. *Phys. Chem. Chem. Phys.*, 15, 20921, 2013.
90. Nakahara, K., Iwasa, S., Satoh, M., Morioka, Y., Iriyama, J., Suguro, M., Hasegawa, E., Rechargeable batteries with organic radical cathodes. *Chem. Phys. Lett.*, 359, 351, 2002.
91. Koshika, K., Sano, N., Oyaizu, K., Nishide, H., An ultrafast chargeable polymer electrode based on the combination of nitroxide radical and aqueous electrolyte. *Chem. Commun.*, 836, 2009.
92. Huang, J., Cheng, L., Assary, R.S., Wang, P., Xue, Z., Burrell, A.K., Curtiss, L.A., Zhang, L., Liquid catholyte molecules for nonaqueous redox flow batteries. *Adv. Energy Mater.*, 5, 1401782, 2015.
93. Zhang, S., Li, X., Chu, D., An organic electroactive material for flow batteries. *Electrochim. Acta*, 190, 737, 2016.
94. Goodenough, J.B., Kim, Y., Challenges for rechargeable batteries. *J. Power Sources*, 196, 6688, 2011.
95. Lu, Y.H., Goodenough, J.B., Rechargeable alkali-ion cathode-flow battery. *J. Mater. Chem.*, 21, 10113, 2011.
96. Lu, Y.H., Goodenough, J.B., Kim, Y., Aqueous cathode for next-generation alkali-ion batteries. *J. Am. Chem. Soc.*, 133, 5756, 2011.
97. Wang, Y.R., Wang, Y.G., Zhou, H.S., A Li–liquid cathode battery based on a hybrid electrolyte. *ChemSusChem*, 4, 1087, 2011.
98. Kreutzer, H., Yarlagadda, V., Nguyen., T.V., Performance evaluation of a regenerative hydrogen-bromine fuel cell. *J. Electrochem. Soc.*, 159, F331, 2012.
99. Braff, W.A., Bazant, M.Z., Buie, C.R., Membrane-less hydrogen bromine flow battery. *Nat. Commun.*, 4, 2346, 2013.
100. Brushett, F.R., Vaughey, J.T., Jansen, A.N., An all-organic non-aqueous lithium-ion redox flow battery. *Adv. Energy Mater.*, 2, 1390, 2012.
101. Su, L., Ferrandon, M., Kowalski, J.A., Vaughey, J.T., Brushett, F.R., Electrolyte development for non-aqueous redox flow batteries using a high-throughput screening platform. *J. Electrochem. Soc.*, 161, A1905, 2014.
102. Andrews, J., Mohammadi, S.S., Towards a 'proton flow battery': investigation of a reversible PEM fuel cell with integrated metal-hydride hydrogen storage. *Int. J. Hydrogen Energy*, 39, 1740, 2014.
103. R.F. Savinell, J.S. Wainright, Iron flow batteries, US Patent 9559375B2, assigned to Case Western Reserve University, 2017.

104. Petek, T.J., Hoyt, N.C., Savinell, R.F., Wainright, J.S., Slurry electrodes for iron plating in an all-iron flow battery. *J. Power Sources*, 294, 620, 2015.
105. Tucker, M.C., Phillips, A., Weber, A.Z., All-iron redox flow battery tailored for off-grid portable application. *ChemSusChem*, 8, 3996, 2015.
106. Alotto, P., Guarnieri, M., Moro, F., Redox flow batteries for the storage of renewable energy: a review. *Renew. Sustain. Energy Rev.*, 29, 325, 2014.

3

Zinc Hybrid Redox Flow Batteries

Srijita Basumallick

PG-Department of Chemistry, Asutosh College under Calcutta University, Kolkata, India

Abstract

In this chapter, an attempt has been made to understand the importance of a Zinc–bromine flow battery as an efficient energy storage material. Elaborate discussions have been made about dendrite formation onto Zn surface during the charging of the battery. Possible use of nano Zn deposited onto carbon felt material as electrode replacing metallic Zn, its merits and demerits have been incorporated. The use of surfactants in electrolytes and their role in the prevention of electrode corrosion and dendrite formation have been explained.

Keywords: Zinc–bromine flow battery, dendrite formation, hydrogen evolution reaction, diffusion-limiting current, pourbaix diagram, quaternary ammonium salts, critical current, proton selective membrane

3.1 Introduction

In recent years, the use of reversible energy storage systems has increased manyfold. Keeping in mind the development of technologies for tapping renewable energy sources like solar, geothermal, wind energy and many other renewable energies, their storage is equally important. It is difficult to get a steady power supply with usable fixed voltage from the energy store. Battery technology acts as an efficient energy storehouse and it is important to make use of these renewable energies. Batteries are mainly of two types. A primary battery, once discharged, cannot be recharged, but a secondary battery can be recharged after discharge. Thus, the secondary battery is reusable. Among several such secondary batteries, only

Email: srijitabasumallick@gmail.com

Inamuddin and Tariq Altalhi (eds.) Advanced Redox Flow Technology, (83–90) © 2024 Scrivener Publishing LLC

a few have a long cycle life. Several bottlenecks associated with the development of efficient batteries can be minimised by the use of redox flow battery technology. Size as well as current density and total energy can be controlled to our desired values. The depth of self-discharge is also much less compared to conventional batteries. A zinc-bromine redox flow battery has some attractive features like moderately high voltage with low-cost electrode materials. Like other secondary batteries during discharge Zn(s) ->Zn^{2+} (aq) + 2e occurs at the anode and Br_2 (aq) + 2e ->2Br^-(aq) in the cathode. A reverse reaction takes place during charging. The anode and cathode compartments are separated by membranes. Bromine and zinc salt solution is stored externally and circulated by a pump as shown in Figure 3.1. Electrolyte pH plays an important role. Overall conductivity, solubility, and electrode potential vary drastically with the change in pH. Pourbaix diagram of zinc [1, 2] helps understand the fate of zinc in different pH environments. Simultaneously side reactions get affected by different pH. At low pH water dissociation into oxygen and hydrogen gas evolution is unlikely to take place [3, 4]. At low pH hydrogen evolution, the reaction can become a competitive reaction at the cathode during charging. Free proton can take an electron from zinc deposited on the electrode while charged followed by corrosion of the electrode. Acetate ions are introduced to act as a buffering agent and take up extra protons. A zinc-bromide redox flow battery acts in a narrow range of pH from 1 to 3.5 [5]. At very high pH zinc hydroxide precipitation takes place which is insulating and does not get reduced to Zn(0) and current efficiency decreases. Proton migration through proton selective membrane is important to keep electro-neutrality for smooth zinc deposition. Not only hydrogen evolution reaction is a threat; any oxidized form of impurities can react with zinc. Cathodic protection with easily oxidizable metals present within zinc electrode reduces this situation. Especially, bromine gas can migrate and react with Zn(0), reducing current efficiency. Here ion-selective membranes prevent bromine gas migration. Corrosion and dissolution of battery materials other than zinc and bromine are also important. So carbon, graphite as well as other polymer materials are used as current collector materials. However, at high positive potential carbon may undergo oxidation producing carbon

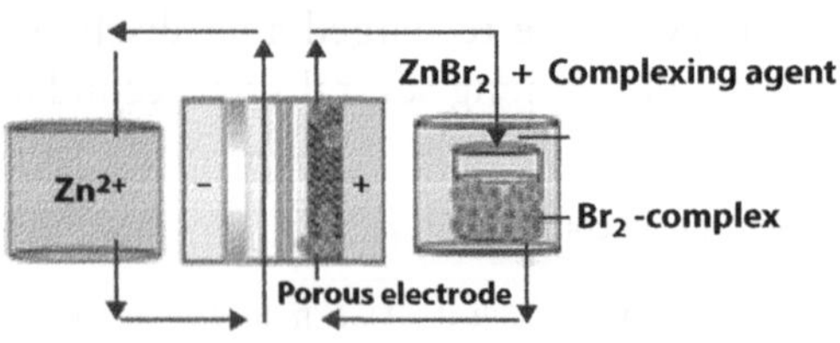

Figure 3.1 Schematic diagram of zinc-bromine redox flow battery.

dioxide. Besides such corrosion at high potential, the advantage of the use of carbon-based materials is due to their light weight. Sometimes conducting polymer binders are used to construct battery materials.

3.2 Zn Electrode and Dendrite Formation

Dendrite formation is a major problem with a metal ion reversible electrode. During charging, uniform deposition of metal on the electrode surface does not occur. Often it has been found that deposited metal crystal grows in a vertical direction, forming a dendrite-like structure. Deposition of Zn onto the Zn surface during charging forms dendrites. It is important to note that at low electrode potential deposition of zinc happens by the charge transfer process. At a little higher potential it becomes a diffusion-limiting current mechanism. At very high potential it is associated with hydrogen evolution reaction (HER). It is interesting to note that the redox potential of Zn /Zn^{2+} may vary slightly depending on the anion present in the solution as well as the pH of the medium. For example, sulphate ion increases the reduction potential of Zn/Zn^{2+}, as zinc sulphate has strong ionic interaction. Whereas chloride and acetate ions maintain reduction potential at a lower value, at a very high potential, hydrogen evolution reaction becomes predominant. Electrode over potential plays an important role in electro crystallization. At a critical current and critical over potential dendrite formation is triggered. Also, it is a diffusion-limiting process. An inhomogeneous concentration of Zn^{2+} creates a high concentration pocket increasing diffusion followed by the formation of dendrite as this pocket reaches critical current. By maintaining homogenized electrolyte concentration particularly through flowing electrolytes, as in the case of flow battery, dendrite formation may be under control. At higher over potential nucleation takes place much faster, resulting in fine crystal grains. Once a dendrite is formed the over potential change takes place together with that high dendrite tip current density helps in the formation of longer dendrites much faster. Dendrite formation happens through several steps. Ion gets absorbed on the electrode followed by reduction and surface diffusion to the nucleation site, then come crystal growth and dendrite formation. At very low over the potential layered boulder and mossy structures are formed, whereas at higher over potential dendrites are likely to be formed. As such dendrite formation doesn't affect the efficiency but it can pierce through the membrane, creating blockage as well as a short circuit. Corrosion of zinc electrodes is another problem. At very low pH hydrogen evolution takes place. Only at high pH (between 11-12) does a zinc oxide protective film form on the electrode. In this case, an electron-taking proton is replaced by dissolved oxygen. Corrosion is influenced by the crystal structure and packing pattern of the electrode surface; finer grain size reduces corrosion. Organic additives slow overall deposition with betterment in morphological character. This reduces corrosion. Some surfactants inhibit the corrosion of the zinc surface. This completely inhibits corrosion. Absorbed additives decrease active sites on to electrode surface. Half-cell reaction associated with that electrode gets affected as well. Several factors affect zinc dendrite formation, such as electrolyte flow rate, electrode material choice, current density, operating temperature, charging rate, and ionic strength. Current pulse can reduce dendrite formation. With the increase in the diffusion of Zn^{2+} ion with each pulse direction of dendrite formation may also change. Here purities of electrode and electrolyte also play an important role in electro crystallization as well as change in over potential. Also, the bromine electrode is highly polarisable, which reduces power density. Since the activity of Br_2/Br^- is much less than zinc, the Zn^{2+}/Zn electrode is modified to increase the activity to increase cell efficiency. Mainly carbon-based electrodes with SWNT [6] and mesostructure [7] are used to increase activity. But exchange current density on carbon-based electrodes are poor. Carbon felt and graphite has higher charge transfer [8] efficiency. A good electrode material should contain a higher surface area and high reaction rate.

3.3 The Electrolyte

The addition of surfactant to the electrolyte solution plays an important role in covering the electrode surface and hinders zinc dendrite growth [9-13]. It is already known that higher over-potential leads to fast nucleation. The addition of long-chain surfactant increases over potential by selectively absorbing onto the surface of the electrode. This prevents grain growth and stimulated nucleation. Electrolyte additives can interrupt conductivity and diffusion rate followed by dendrite growth. The inhomogeneous electrode surface plays an important role in dendrite formation. Instead of using zinc metal as an electrode, Zn deposited on graphite felt is seen to reduce dendrite formation. This is not only because of the high electro-active area for plating. It inhibited voltage hysteresis, improving cycling stability. Inhibiting a steady diffusion current can help solve the dendrite formation problem. A porous electrode with a high surface area and low charge density is a better choice. Here charge transfer control charging-discharging process brings improvement in the dendrite formation problem. Compared to carbon nanotube grafted with zinc electrode zinc oxide coated electrode is a smart way to inhibit dendrite in electrode surface. Zinc oxide is highly insulating. Crystal defects on the zinc oxide layer on the electrode surface create a porous site in it. Zinc can undergo redox reactions through these pores as usual. It is important to mention that in the case of graphite felt grafted with zinc, a single vacancy in the carbon site helps uniformly absorb zinc. This also helps in reducing dendrite formation. Not only the modification of electrodes on the zinc side is necessary. If the activity of Br_2/Br^- is increased the overall kinetics of reactions will increase. This will also lead to higher current density followed by retarding Zn dendrite formation [14]. Ion-selective electrodes facilitate inhibition in dendrite formation. An increase in temperature increases diffusion, and as a result, the dendrite formation gets inhibited. Cationic surfactants like tetrabutylammonium chloride (TBACl) and cetyltrimethylammonium bromide (CTAB) are often used to hinder dendrite growth as well as refine grain boundaries. But cationic surfactants are not stable as anionic ones. Perfluorosurfactants [15] are highly stable anionic surfactants but simultaneously costly. Another interesting category of molecules with highly specific interaction on electrode surfaces but smaller in size is known as levellers. These levellers bind more in the peak positions compared to valleys or recessed areas. As a result, electro deposition or crystallization happen more in recessed areas, truly levelling the electrode surface. For that optimal concentration of leveller needs to be used. Also, some inorganic materials like indium oxide show high hydrogen evolution over potential. Zinc bromide is used as a primary electrolyte (1-4 M) [5]. Chloride-based electrolytes improve the conductivity as well as the energy efficiency of the battery. Methansulfonic acid reduces internal resistance followed by a fast oxidation-reduction reaction [16] in the battery. A bromine sequestration agent like 1-Ethyl-1-methylpyrrolidinium bromide [C2MPyrr]Br helps reduce migration through the membrane but it forms different phase with bromine-reducing bromine reduction kinetics [17].

3.4 Effect of Temperature

To increase the total energy stored in a zinc bromide flow battery, capacity storage as well as the concentration of electrolytes needs to be high. At high concentration of bromine, a diffusion problem and a short circuit problem become significant. Any inhomogeneous bromine reduction triggers inhomogeneity in zinc deposit followed by dendrite formation. At relatively higher temperature bromine side may form inhomogeneity. During charging when temperature increases bromine vapour liberated can cause corrosion and toxic gas liberated cause safety hazards. Quaternary ammonium salts [18] like Tetra-n-butyl ammonium (TBA) is used as electrolyte. Bromine forms complexes with these quaternary ammonium salts followed by a reduction in volatility as well as bromine diffusion towards an object side is reduced. During discharge bromine complexes slowly break up automatically. So this slow release helps in the prevention of zinc dendrite formation. The temperature affects electrode potential, electrolyte solubility, electrolyte resistance, reaction kinetics, and diffusion coefficient of active species. Independent of the pH of the electrolyte solution, coulombic efficiency is seen to increase. But at high temperatures hydrogen adsorbed on the electrode is removed, followed by dissolved oxygen-mediated corrosion. At low-temperature electrolytes, resistivity exceeds the effect from bromine crossover discharge. Thus at low temperatures voltage efficiency and coulombic efficiency show the opposite trend. A similar model in high temperature shows coulombic efficiency beat voltage efficiency. The overall temperature effect is negligible under a perfect situation of a flow system.

3.5 The Membrane

Bromine can diffuse to the zinc side, forming a short-circuit current. To prevent bromine diffusion ion-selective membrane is used. But diffusion of cations, protons and water needs to percolate inside the membrane hole. This needs membrane pore diameter to be optimized. Nafion membrane is most suitable for this purpose. Bromine cannot pass through this membrane; this may be due to the formation of Br_3^- and Br_5^-. Adding extra doping iron can improve the membrane. To keep the degradation of the membrane and battery at bay polyolefin film is used as a membrane. A membrane made of cheaper materials like polyethene and silica gel with low resistance is perfect for this purpose. Anionic impregnation reduces bromine leakage through the membrane. The membrane should be low cost, high cycle life, with low resistance and high water contents. The membrane can be of two types: 1) The micro-porous membrane, which is less efficient, and 2) Ion exchange membrane, which is costly. Quaternizedpolysulfone composite membranes (QNPSU) [19] are more effective as membrane material compared to Nafion membranes. Activated carbon-coated membranes proposed by Zang *et al.* [20] have a higher surface area and absorb bromine and bromide ion that helps in the reduction of over potential and interfacial resistance.

3.6 Hydrogen Evolution Reaction

As already discussed, hydrogen evolution reaction can compete with zinc deposition reaction. Several additives are used to suppress the hydrogen evolution reaction like CTAB (Cetyl tri methyl ammonium Bromide). This can be used to increase the coulombic efficiency of battery performance. Also, succinic acid and tartaric acid can be used in place of CTAB. The major problem associated with surfactants is that these additives are of high molecular weight and can reduce the overall conductivity of the battery.

3.7 Conclusion

In this chapter, we have discussed some important features of a zinc-bromine flow battery. The major problem of dendrite formation onto the Zn surface has been elaborately discussed. The merits and demerits of the use of surfactants in electrolyte solutions have been discussed. Instead of using Zn metal as an electrode, Zn is deposited in nano form onto carbon felt material, and its advantages and disadvantages have been discussed.

References

1. Beverskog, B., Puigdomenech, I., Revised pourbaix diagrams for zinc at 25–300 °C. *Corros. Sci.*, *39*, 107, 1997.
2. Hoang, T. K. A., Doan, T. N. L., Sun, K. E. K., Chen, P., Corrosion chemistry and protection of zinc & zinc alloys by polymer-containing materials for potential use in rechargeable aqueous batteries. *RSC. Adv.*, *5*, 41677, 2015.
3. Leung, P. K., Ponce-de-León, C., Low, C. T. J., Shah, A. A., Walsh, F. C., Characterization of a zinc–cerium flow battery. *J. Power Sources*, *196*, 5174, 2011.
4. Giménez, S. (Ed), Bisquert, J. (Ed.), *Photoelectrochemical Solar Fuel Production*, Springer Cham, Switzerland, 2016.
5. Rajarathnam, G and Vassallo A., *The Zinc/Bromine Flow Battery: Materials Challenges and Practical Solutions for Technology Advancement*, pp. 45-46, Springer, Singapore, 2016.
6. Munaiah, Y., Dheenadayalan, S., Ragupathy, P., Pillai, V. K., High Performance Carbon Nanotube Based Electrodes for Zinc Bromine Redox Flow Batteries. *ECS. J. Solid. State. Sci. Technol.*, *2*, M3182, 2013.
7. Wang, C., Li, X., Xi, X., Zhou, W., Lai, Q., Zhang, H., Bimodal highly ordered mesostructure carbon with high activity for Br2/Br– redox couple in bromine based batteries. *Nano. Energy.*, *21*, 217, 2016.

8. Pandolfo, A. G., Hollenkamp, A. F., Carbon properties and their role in supercapacitors. *J. Power Sources, 157,* 11, 2006.
9. Gomes, A., da Silva Pereira, M. I., Zn electrodeposition in the presence of surfactants: Part I. Voltammetric and structural studies. *Electrochim. Acta., 52,* 863, 2006.
10. Alfantazi, A. M., Dreisinger, D. B., An investigation on the effects of ortho-phenylene diamine and sodium lignin sulfonate on zinc electrowinning from industrial electrolyte. *Hydrometallurgy, 69,* 99, 2003.
11. Gomes, A., Viana, A. S., da Silva Pereira, M. I., Potentiostatic and AFM Morphological Studies of Zn Electrodeposition in the Presence of Surfactants. *J. Electrochem. Soc., 154,* D452, 2007.
12. Pan, J., Wen, Y., Cheng, J., Pan, J., Bai, Z., Yang, Y., Zinc deposition and dissolution in sulfuric acid onto a graphite–resin composite electrode as the negative electrode reactions in acidic zinc-based redox flow batteries. *J. Appl. Electrochem., 43,* 541, 2013.
13. Leung, P. K., Ponce-de-León, C., Low, C. T. J., Walsh, F. C., Zinc deposition and dissolution in methanesulfonic acid onto a carbon composite electrode as the negative electrode reactions in a hybrid redox flow battery. *Electrochim. Acta., 56,* 6536, 2011.
14. Yang, H. S., Park, J. H., Ra, H. W., Jin, C.-S., Yang, J. H., Critical rate of electrolyte circulation for preventing zinc dendrite formation in a zinc–bromine redox flow battery. *J. Power Sources, 325,* 446, 2016.
15. Trejo, G., Ortega B, R., Meas, Y., Ozil, P., Chainet, E., Nguyen, B., Nucleation and Growth of Zinc from Chloride Concentrated Solutions. *J. Electrochem. Soc., 145,* 4090, 1998.
16. Wu, M. C., Zhao, T. S., Wei, L., Jiang, H. R., Zhang, R. H., Improved electrolyte for zinc-bromine flow batteries. *J. Power Sources, 384,* 232, 2018.
17. Schneider, M., Rajarathnam, G. P., Easton, M. E., Masters, A. F., Maschmeyer, T., Vassallo, Anthony M., The influence of novel bromine sequestration agents on zinc/bromine flow battery performance. *RSC. Adv., 6,* 110548, 2016.
18. *EPRI-DOE Handbook of Energy Storage for Transmission and Distribution Applications,* EPRI- 1001834, 2003.
19. Li, M., Su, H., Qiu, Q., Zhao, G., Sun, Y., Song, W., A Quaternized Polysulfone Membrane for Zinc-Bromine Redox Flow Battery. *J. Chem., 2014,* 321629, 2014.
20. Zhang, L., Zhang, H., Lai, Q., Li, X., Cheng, Y., Development of carbon coated membrane for zinc/bromine flow battery with high power density. *J. Power Sources, 227,* 41, 2013.

4

Zinc-Bromine Hybrid Redox Flow Batteries

M. Ramesh[1]*, J. Maniraj[1], D. Santhosh Kumar[2] and S. L. Pradeep Kumar[2]

[1]Department of Mechanical Engineering, KIT-Kalaignarkarunanidhi Institute of Technology, Coimbatore, Tamil Nadu, India
[2]Department of Aeronautical Engineering, KIT-Kalaignarkarunanidhi Institute of Technology, Coimbatore, Tamil Nadu, India

Abstract
Biomass-derived activated carbon is in high demand as electrodes and electrocatalysts due to its ease of availability, excellent chemical stability, higher surface contact, and cheap and environmentally friendly nature. The efficiency of the Zn-Br redox flow battery (ZBRFB) is inversely proportional to the positive electrode's surface characteristics. The total performance of the ZBRFB system depends critically on the bromine/bromide redox pair's reversibility. RFB has lower energy density than lithium-ion batteries owing to its low output voltage. The maximum discharge capacity and cycle stability of ZBRFB will be increased due to the uniform zinc distribution ability.

***Keywords*:** Zn-Br, redox flow batteries (RFB), lithium-ion batteries, bromide batteries

4.1 Introduction

Renewable resources became necessary when natural resources were continuously depleted. Energy storage technology is far superior to any other technology in its ability to allow for the flexible use of intermittent resources [1]. This is supported by the fact that RFBs such as zinc bromide and vanadium batteries are widely used in the commercial sector around

**Corresponding author*: mramesh97@gmail.com

Inamuddin and Tariq Altalhi (eds.) Advanced Redox Flow Technology, (91–106) © 2024 Scrivener Publishing LLC

the world [2]. As of late, research on aqueous redox flow Zn: I-Br-based battery has shown a higher energy density of 101 $WhKg^{-1}$. The high redox potential of the Br-redox pair is where our real focus lies in the reversibility of iodide. Zn-Br is also the subject of intense research due to its high theoretical energy density and increased cell voltage [3]. To achieve widespread commercialization of the Zn–Br flow battery (ZBFB), however, it is necessary to resolve crucial concerns such as low reliability, poor energy efficiency, bromine smoke evolution, and poor kinetics. The Br ions are oxidized during charging to generate elemental bromine at the positive and Zn^{2+} ions at the negative compartment. Discharge involves the opposite reaction taking place at each electrode [4].

The ZBFB and the VRFB, which have been extensively investigated, have inherent benefits and drawbacks. The main benefit of the VRFB is that it only uses one redox species, which theoretically allows for infinite storage. Even though, pH environment is usually necessary to stabilize the salts. This might cause a capacity fade due to parasitic reactions like the hydrogen evolution reaction. On the other hand, ZBFB has a smaller parasitic reaction rate and a more significant cell operating voltage [5]. ZnBr is highly soluble in water (528 g/100 ml of H_2O), and ZBFB can produce high specific energy densities. The battery's capacity is set by the amount of zinc metal put on the negative electrode. Consequently, unlike VRFB, power and capacity in ZBFB are not fully separated. Restricting the elemental Bromide developed in cell charging is the major issue with ZBFB [6]. Figure 4.1 shows the evolution of rechargeable zinc batteries. These cells have a lengthy history and, in certain instances, exceptional dependability. Classical cell batteries are well known. When coming to the topic of redox flow batteries, 40 years of research and commercialization have been done in zinc-bromine-type batteries. The zinc-air and zinc-cerium redox flow batteries are still under examination. The zinc-nickel redox flow battery was established into a commercial one, while zinc-iron or zinc polyiodide is under development [7].

4.2 Electro-Chemical Energy Storage

Zinc batteries, which can be recharged, are a great option for storing energy because they can feed energy back into the system for several days. The potential and benefits of zinc in stationary storage are being demonstrated by zinc batteries, which come in various chemistries and designs and are progressively meeting the needs of the residential, commercial, and microgrid energy storage markets [8]. The electrochemical system was used to

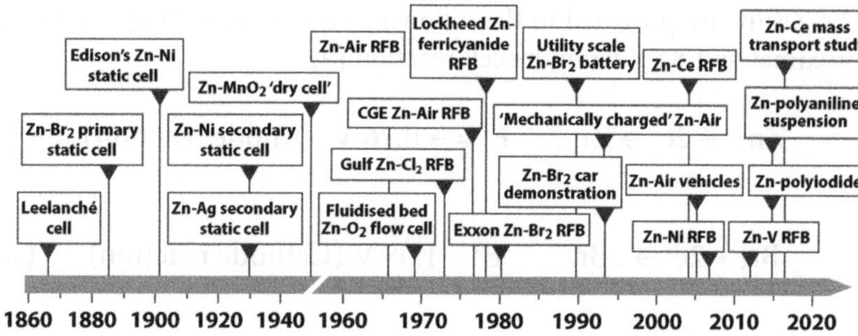

Figure 4.1 Evolution of Zinc rechargeable flow batteries [7].

visualize high-energy density which is very important for storing electricity from natural sources.

$$Br_2 + 2e^- \leftrightarrow 2Br^- \qquad E^0 = +1.08 \text{ V vs SHE (positive electrode)} \quad (4.1)$$

$$Zn \leftrightarrow Zn^{2+} + 2e^- \qquad E^0 = -0.76 \text{ V vs SHE (negative electrode)} \quad (4.2)$$

$$Zn + Br_2 \leftrightarrow Zn^{2+} + 2Br^- Ecell = 1.84 \text{ V (Net reaction)} \quad (4.3)$$

4.3 Redox Flow Batteries

Energy efficiency, safety, dependability, and affordability are all essential features of a great energy storage system. The RFB is a popular option in this field because of its many advantages such as flexible design and decoupled control. Generally, RFBs constitute an ion exchange membrane and microporous separator, an electrode, and an electrolyte. Several RFBs, including vanadium, ZrBr metals, Fe-Cr, and Fe-Va, have been successfully validated so far. ZBFBs are the most practical option because of their high energy density (70 Wh/kg) and ready availability of raw materials. Zinc is deposited on the negative electrode terminal by cathodoluminescence; a poly-bromide phase is synthesized at the positive electrode via anodo luminescence during the charging process of ZBFB. The reaction was reversed during the discharge phase. The electrochemical reactions in the ZBFB are provided in the following equations, getting a voltage of approximately 1.84 V from the cell reaction [9].

$$Zn^{2+} + 2e^{-} \rightarrow Zn \qquad E^{0} = -\,0.76\text{ V (Anode reaction)} \tag{4.4}$$

$$Br_2 + 2e^{-} \rightarrow 2Br^{-} \qquad E^{0} = 1.08\text{ V (Cathode reaction)} \tag{4.5}$$

$$Zn + Br_2 \leftrightarrow Zn^{2+} + 2Br^{-} E_{cell} = 1.84\text{ V (Net reaction)} \tag{4.6}$$

Electrode material selection plays a crucial role in RFB. The vast operating potential window of most RFBs necessitates that all components be highly chemically stable. Numerous studies have shown the effectiveness of carbon-based materials as RFB electrodes. However, their stamina, power density, and rate capacity are subpar. Therefore, the primary approaches to achieving highly effective energy storage are the development of suitable electro-catalysts and electrode performances boosted by modifying the surface of carbonaceous materials [10].

A thick graphite plate with a carbon or graphite-felt active electrode is typical of a flow battery. This thick graphite electrode's low catalytic activity is notably pronounced concerning the Br_2/Br redox pair, resulting in pronounced polarisation. In addition, the current density in the system was reduced when thick carbon/graphite electrodes were used. Therefore, 20-30 mA/cm^2 is a standard range for evaluating the Zr-Br redox flow battery (ZBRFB) system. Significant voltage drop is observed in the cells at this current rate, which has a devastating effect on the flow cell's performance and, particularly, its voltaic efficiency [8]. Highly conductive and lightweight electrode materials are highly valued for reducing Ohmic loss and overall stack weight. Carbonaceous materials have been the subject of a plethora of research trying to lower internal resistance and boost electrocatalytic activity through various surface treatment techniques such as heteroatom doping, acid treatment, surface oxidation, and surface functionalization [6]. It is helpful to modify the surface of electrodes by adding hydroxyl, epoxy, carboxyl, and carbonyl which tend to improve electrode kinetic processes [11]. Figure 4.2 shows, among the aqueous system, several RFBs reveal the highest cell potential, mostly at least 400 mV greater than all vanadium cells. Zn-based RFBs compromise the relatively low charging efficiency. It is critical to control the position of the zinc electrode

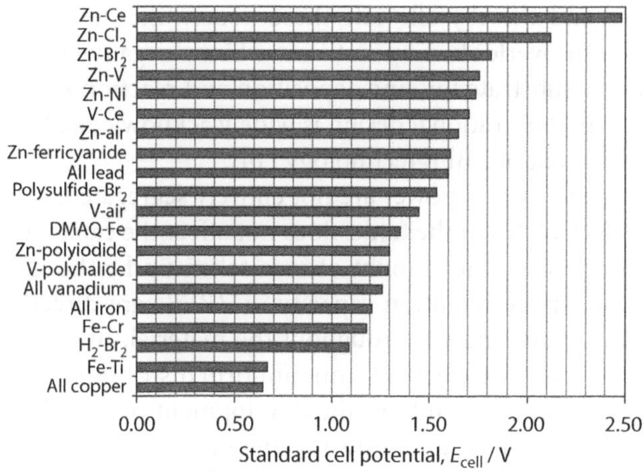

Figure 4.2 The standard cell potential of selected aqueous RFBs [7].

to affect compact, dendrite-free deposits, To prevent the short circuit, the bipolar electrode (negative terminal) is coated uniformly. Over a repeated battery cycle, the oxidation of metallic deposits is required to prevent accumulation [7].

4.4 Zinc/Bromine Flow Batteries

RFBs are becoming increasingly popular for energy storage owing to their high proficiency, excellent reliability, easy design, rapid reaction, and fatigue life. One of the most cutting-edge types of redox flow batteries is the ZBFB. Redox reactions occur in the pile of two electrodes separated by an ion-exchange separator. Chemically active electrolytes were kept in a storage tank [12]. Bromide species (Br_3 or Br_5) will be produced during the charging process of zinc and bromide. Despite the substantial progress made in ZBFB development, several serious issues still hinder ZBFBs from being widely deployed in the industry. Concerns about the presence of the deadly charge product Br_2 are brought up, for instance, by the high cell polarisation generated by the slow kinetics of the Br_2/Br redox reaction at the positive electrode. Low ionic conductivity and high ohmic resistance are produced by the $ZnBr_2$ acidic solution [13]. The positive electrode has been modified with various nanomaterials to increase electrochemical performance and thereby alleviate these issues. The downside, unfortunately, has been overlooked up until now. Since negative electrodes have a dual role as a substrate for zinc deposition and reaction sites for the Zn/Zn^{2+} redox pair, their rational design is widely acknowledged to contribute to the development of ZBFB technology [14].

Therefore, energy and power are not entirely segregated. Therefore, the number of electrolytes in the tanks does not affect the capacities of ZBFBs at a given stack size. The amount of area taken up by the negative electrode and the charge retention capacity of ZBFBs are directly related to how well zinc is distributed throughout the material. In addition, ZBFBs Zn-Cl_2, Zn-I_2, and Zn-Ce benefit from uniform distribution since it avoids localized zinc nucleation, inhibit the development of zinc dendrites, and enhance cycling stability. In contrast to the metallic plate electrodes used in the zinc deposition process of the electro-plating industry, graphite and carbon felt are commonly utilized as electrodes in flow batteries. This is due to the many benefits of these materials, including their low cost, high specific surface area, high porosity, and superior chemical stability. That's why it's just as crucial to deposit zinc inside the pores of the porous electrode as it is to place it on its exterior [15]. The scheme of Zinc-bromine batteries is provided in Figure 4.3 [16].

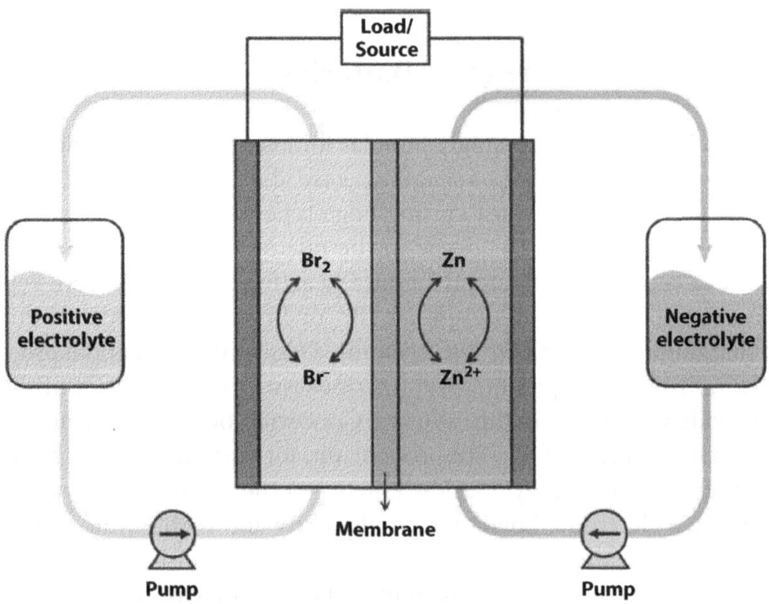

Figure 4.3 Schematic of Zinc-bromine batteries [16].

4.5 Types of Zinc-Based Hybrid Flow Batteries

4.5.1 Zinc-Sulphur (Zn–S) Hybrid Battery

Zinc metal has an effective redox potential. Its reliability, safety, renewability, and cost-effectiveness are all priorities. It will be used as an anode with a high specific capacity. But the anode, such as manganese and vanadium oxides, results in a low specific capacity. Sulfur is a commercially viable cathode material due to its natural availability, low cost, non-toxicity, and extremely high theoretical specific capacity. Four-electron transfer process on the sulfur cathode produces extremely high specific capacity and energy density. The high reversible discharge capacity of the Zn-S battery is 2063 $mAhgS^{-1}$. Therefore, the Zn-S hybrid battery is a secure, adaptable energy storage technology with various applications in the development of flexible electronic systems [3].

4.5.2 Zinc-Nickel (Zn/Ni) Batteries

Rechargeable zinc-nickel (Zn/Ni) batteries have high demand due to their high output voltage, affordability, eco-friendliness, etc. Cycle stability is low, and the cathode material's capacity does not match the anode. For this, nickel selenides with outstanding electrical conductivity and redox states, such as NiSe2 and NiSe, were employed as cathode materials [6]. The activation and degradation of the NiSe-2 h electrode were slowed by adding potassium ferricyanide to the KOH electrolyte. The NiSe/Zn-KFC battery has 210 cycles of peak capacity, which is more than the NiSe/Zn-U battery (124 cycles).

4.5.3 Zinc-Sodium Hybrid Ion Batteries (ZSHIBs)

Standard zinc batteries cause significant resource waste and pollution to the environment as they are hardly recyclable. The efficient route to solve these problems is to implement ZSHIBs in which metallic zinc is an anode and sodium is a cathode. The Na^+ ions reduce the zinc dendrites to a great extent. Sodium-based metal-ion batteries are cost-effective and have a long cycling life, high energy density, and high rate capability for large-scale storage systems. But the level of safety is poor for specialized equipment like flexible watches, clinical wearable devices, and unsecured test equipment, which require a stable power supply [3].

4.5.4 Zn-Ion Batteries (ZIBs)

Zinc ion batteries are the original aqueous hybrid battery that has gained popularity owing to their merits cost-effectiveness, non-flammability, and environmental friendliness. The matrix contains crystalline silver nanoparticles and sloppily packed polyoxometallates (POM) clusters which assimilate electrochemical and polyoxoanionic redox reactions with passable electron diffusion kinetics. POM-based electrode materials for Zinc-based batteries have high recycling stability, high durability, and a stable energy storage capacity [17].

4.6 Electrochemistry of Zinc/Bromine Deposition

4.6.1 Electrochemical Performance

The performance of Zn-Br2 is significantly influenced by the flow rate of the electrolyte. The most favorable flow rate for ZBFB performance is the

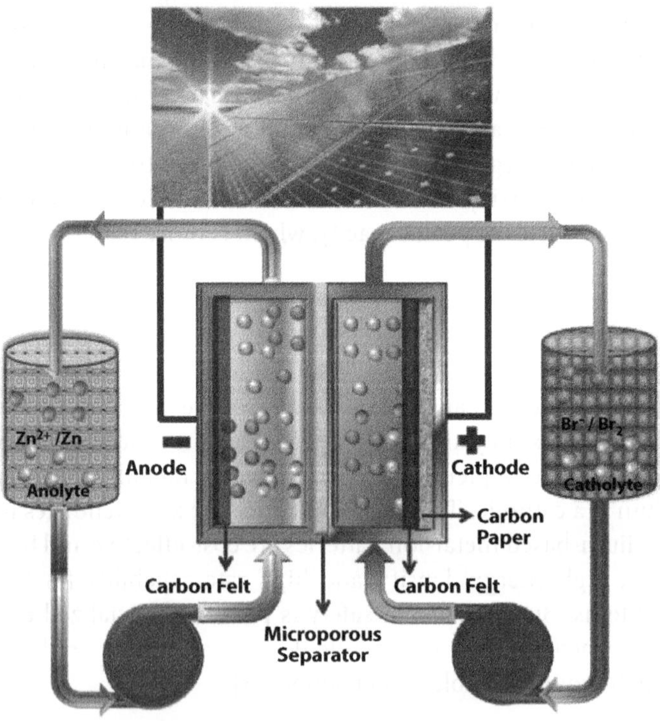

Figure 4.4 Operating principle of zinc-bromine batteries [10].

asymmetric electrolyte flow rate condition. At 20 and 30 ml per minute electrolyte flow rates, the cell functioned flawlessly. Improper zinc plating on the electrode surface is caused by the increasing electrolyte flow rate. High current densities resulted in a shorter discharging time. Meanwhile, as the current density increases, the cell becomes more polarized. The per-chloric acid concentration and efficient electrolyte flow rate condition were the two most promising ways to increase the Br/Br2 redox kinetics in a Zn-Br2 RFB [2]. The working principle of the ZBRFB single-flow cell is described in Figure 4.4. The setup consists of a positive electrode of dimension 3 cm x 3 cm where the carbon paper of thickness 0.28 was employed in between the graphite plate 2.5 mm grooved on the surface and the carbon felt of 3 mm thick, carbon paper was placed. Microporous membrane separated the positive and negative sides to study the galvanostatic charge-discharge performance Arbin – BT 2000 instrument is used. The fabricated cell wall is tested at constant current and discharges at various current densities [10].

4.6.2 Reduction of Dendrite Deposition

A specific threshold of cell potential is required to develop zinc dendrites—the intensity of current and electrolyte concentration near the site of dendrite formation. Dendrites originate at surface irregularities with high local current density and grow to varying heights. Dendrites grow on the carbon cloth electrode as zinc is deposited. Undoubtedly, adding organic chemicals reduces the dendrite trunk height and the number of dendrite branches. Dendrite formation was accompanied by a side reaction—hydrogen evolution—which the organic chemical additives can mitigate. Zn-Br battery has been developed using neutral, alkaline hybrid electrolytes and carbon-manganite as an effective cathode catalyst. Excellent high-rate performance and a high average discharge voltage of 2.15 V are hallmarks of this battery system [8]. The factor that plays a role in Zn deposition is listed in Table 4.1.

4.6.3 Bio-Mass Electrocatalyst

It is common for campuses to collect and process bio-waste for natural fertilizers. Still, sometimes waste is left where it is, accumulating mountains of trash that must be burned or incinerated, potentially polluting the surrounding area [18, 19]. Leaves of the Indian gooseberry tree are used to produce activated carbon by increasing the kinetics of the Br/Br_2 redox reaction at the electrodes in Zn-Br batteries. Thus, the cost-efficient, non-polluted

Table 4.1 Factors that play a role in Zn deposition.

S. No.	Aspects	Effects	Results	Remedies
1	Electrode	Plate electrodes exacerbated dendrite growth because of their small active surface area.	Electrical conductivity, thickness, hydrophilicity, mechanical strength, and chemical stability were decreased.	By using porous electrodes with large surface area.
2	Electrolyte composition	High electrolyte concentration will inhibit Zn2+ ion reduction by reducing its transport to electrodes.	The solubility of zincate ions and electrochemical performance were suppressed.	Use suitable additives, and Zn2+ ion concentrations above 0.4 M indicate successful plating.
3	Flow rate	An electrolyte leak or energy loss will occur if the flow rate is too high.	Decrease in mass transfer rate.	The flow rate must be precisely controlled for optimal performance to strike a balance between Zn dendrites and pressure drop.
4	Current density	High current density proliferates Zn dendrites.	The battery will get failure.	ZFBs need to be operated at a low current density.
5	Temperature	As the temperature drops, ions are deposited on the nucleation site, causing crystallization.	Increased electrolyte viscosity and sluggish ion mobility.	High temperatures prevent Zn dendrite formation.

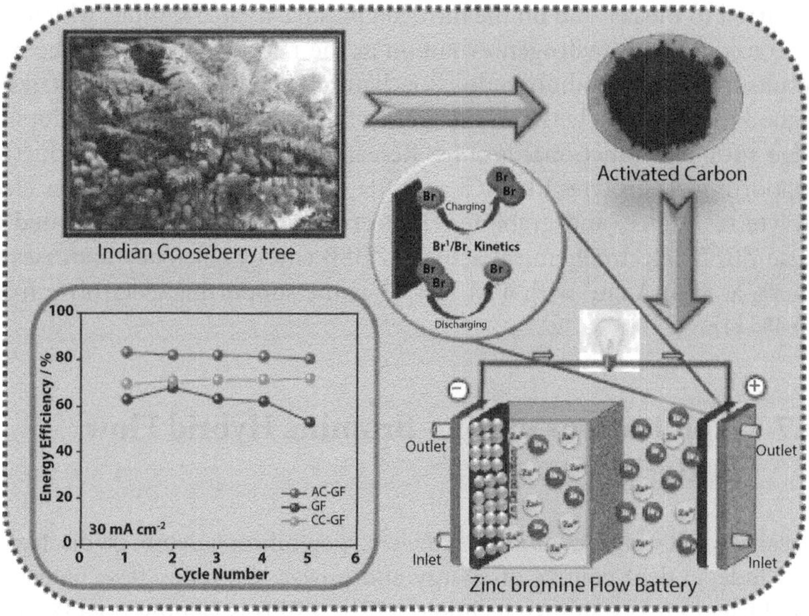

Figure 4.5 Bio-mass extraction process [9].

alternate current derived from the bio-mass acts as a potential electrocatalyst for excellent performance of the batteries (Figure 4.5) [9].

4.6.4 Surface Chemistry

To a large extent, the carbon electrode's surface chemistry determines the positive electrode's electrochemical performance in ZBFB. Implementation is primarily examined by the polymer phase and electrode surface. At high current densities, when the concentration of the polybromide phase is higher, the large number of oxygen functional groups proves particularly useful. The polyhalide ion activity in each phase governs the dominant phase separation process during the electrolytic formation of the bromine-fused salt, which is modeled as a micelle-like process. By converting Br_2 into a condensed, conductive, and chemically stable fluid upon standing, this complexation method can be used for energy storage [20, 21].

4.6.5 Effect of Zinc Utilization

Faradaic efficiency was improved by switching to a graphite plate without an electrolyte quality of 99.9 wt. % of zinc utilization. Capacity fade occurred in the cell run on the nitrogen plasma-treated sample, which can be traced back to hydrogen evolution at the negative electrode. The best results from the vanadium redox flow batteries can be achieved by striking a good compromise between the levels of defects like heteroatom doping, edge sites, and functional groups. Research into highly ionic conductive supporting electrolytes like KCl and NH_4Cl was conducted to lower electrolyte resistance, and graphite-felt electrodes were thermally treated to raise ZBFB's electrochemical activity. ZBFB energy efficiency increases to 74.3% at 40 mA cm^2 with 4 M NH_4Cl as the supporting electrolyte from 60.4% [4].

4.7 Applications of Zinc-Bromine Hybrid Flow Batteries

Because of its high coulombic efficiency, significant depth of discharge, and independent control of energy and power, the redox flow battery is an efficient energy storage technology. The ZBFB is gaining popularity as an alternative to other redox flow battery technologies due to its low cost, long lifetime, and high energy efficiency. With a maximum power input of l/4, the zinc-bromine system can generate a limiting specific energy of 124 Wh/kg. In a normal load-balancing cycle, the battery is charged for six hours and four hours of discharge time at l/2 maximum power, and the zinc-bromine system achieves 75% energy savings at 30 W/kg. Its enhanced specific energy and power can be used for electrochemical applications such as electrode substrates, cell housings, pump casings, reservoirs, etc. [20].

4.8 Future Challenges

4.8.1 Electric Vehicles

Slow recharging times, high costs, improper disposal, materials composition, safety, size, refueling station, and low power density are some problems plaguing energy storage systems, especially in electric vehicles. Power electronics interface, hydrogen viability, battery lifetime, energy management system, etc., are some of the issues that must be overcome in developing fuel cell and battery electric vehicles. Superconductors, supercapacitors, and fuel cells are all crucial to manufacturing energy storage systems. These issues significantly affect the development of energy storage systems, which is particularly problematic for electric vehicles [22, 23].

Table 4.2 Advantages and disadvantages of zinc batteries.

S. No	Advantages	Disadvantages
1	Enhanced specific energy	Circulation and temperature regulation necessitate supplementary systems.
2	Capability for variation in overall system layout	As with all batteries, system design must ensure safety.
3	Environment-friendly, room-temperature functionality	A significant amount of energy is lost when the device is turned off while being charged.
4	Being able to charge quickly	Potentially necessary enhancements to moderate power capacity.
5	Crafted from easily obtainable and inexpensive components	Risk of dendrite formation.

4.8.2 Energy Management

Hybrid energy sources, UCs-FCs, and batteries are just a few of the many devices needed to keep an electric vehicle running, and they all need to be recharged from time to time. Consequently, electric vehicles necessitate an energy management system to control all power sources and ensure a constant power supply to the vehicles' energy storage systems [13].

4.8.3 Size and Cost

Electric storage systems account for nearly one-third of the total production cost of electric vehicles; however, their low price and small size pose significant obstacles to the advancement of electric vehicles [24]. While the overall price of electric storage systems is high due to maintenance, operation, and labor some components are inexpensive and can be swapped out for more expensive ones [2].

4.8.4 Safety Measures

Safety precautions are crucial not only during the manufacturing process but also for the upkeep of the energy storage system [25]. The bromine in an electrolyte is either free bromine or bonded to a completing agent in the aqueous phase. The aqueous electrolyte dissolves any traces of bromine that may be present. Bromine, whether in liquid or gaseous form, is toxic and can cause serious harm if it comes into contact with the skin or is inhaled. However, in its complex state, bromine's chemical reactivity and evaporation rate are drastically diminished compared to pure bromine [26].

4.9 Conclusion

Attention to the safest energy-generating technology has grown in tandem with the proliferation of new and renewable electricity-generation technologies. The energy storage system (ESS) can serve various purposes like balancing fluctuations, stabilizing voltage regulating frequency, lowering emissions, and shaving peaks. The efficiency of ZB cells' current generation is notoriously unstable. This is because bromine reduction is impeded, and the bottom surface of the electrode is deactivated due to poor mixing of the polybromide complex phase at the bottom of the test cell during discharging. Incorporating a few drops of polysorbate solved these problems, improving current efficiency due to better catholyte mixing across the electrode surface. Thus, focusing on reducing the dendrite growth by proper methods will make the zinc-bromine hybrid redox flow batteries attractive with their low costs, long lifetime, and high energy efficiency. Research has to be steered to extend its applications for better batteries for mobile devices.

References

1. Ramesh, M., Ravanan, A. Microwave-assisted inorganic materials for supercapacitors. In: *Inorganic Nanomaterials for Supercapacitor Design*, Inamuddin *et al.* (Eds.) pp. 101-117, 2020.
2. Adith, R.V., Naresh, R.P., Mariyappan, K., Ulaganathan, M., Ragupathy, P. An optimistic approach on flow rate and supporting electrolyte for enhancing the performance characteristics of Zn-Br_2 redox flow battery. *Electrochim. Acta*, 388, 138451, 2021.
3. Dai, C., *et al.* Maximizing energy storage of flexible aqueous batteries through decoupling charge carriers. *Adv. Energy Mater.*, 11(14), 1, 2021.
4. Naresh, R.P. *et al.* Activated carbon-anchored 3D carbon network for bromine activity and its enhanced electrochemical performance in Zn–Br_2 hybrid redox flow battery. *ChemElectroChem*, 6(22), 5688, 2019.
5. Archana, K.S., *et al.* Effect of positive electrode modification on the performance of zinc-bromine redox flow batteries. *J. Energy Storage*, 29, 101462, 2020.

6. Huang, Y., *et al.* Constructing aqueous Zn//Ni hybrid battery with NiSe nanorod array on nickel foam and redox electrolytes for high-performance electrochemical energy storage. *Appl. Surf. Sci.*, 562, 150222, 2021.
7. Arenas, L.F., Loh, A., Trudgeon, D.P., Li, X., de Leon, C.P., Walsh, F.C. The characteristics and performance of hybrid redox flow batteries with zinc negative electrodes for energy storage. *Renew. Sustain. Energy Rev.*, 90, 992, 2018.
8. Park J.H., Steingart, D., Koel, B. Visualizing zinc dendrites in minimal architecture zinc bromine batteries via in-house transmission X-ray microscopy. *Microsc. Microanal.*, 27, 2448, 2021.
9. Suresh, S., Kesavan, T., Munaiah, Y., Arulraj, I., Dheenadayalan, S., Ragupathy, P. Zinc-bromine hybrid flow battery: Effect of zinc utilization and performance characteristics. *RSC Adv.*, 4(71), 37947, 2014.
10. Suresh, S., Ulaganathan, M., Aswathy, R., Ragupathy, P. Enhancement of bromine reversibility using chemically modified electrodes and their applications in zinc bromine hybrid redox flow batteries. *ChemElectroChem*, 5(22), 3411, 2018.
11. Yuan, X., *et al.* An aqueous hybrid zinc-bromine battery with high voltage and energy density. *ChemElectroChem*, 7(7), 1531, 2020.
12. Lee, J.N., Do, E., Kim, Y., Yu, J.S., Kim, K.J. Development of titanium 3D mesh interlayer for enhancing the electrochemical performance of zinc–bromine flow battery. *Sci. Rep.*, 11(1), 1, 2021.
13. Xu, Z., Fan, Q., Li, Y., Wang, J., Lund, P.D. Review of zinc dendrite formation in zinc bromine redox flow battery. *Renew. Sustain. Energy Rev.*, 127(2), 109838, 2020.
14. Mulleriyawage, U.G.K., Shen, W., Hu, C. Battery system selection in DC microgrids for residential applications: An Australian case study. *Proc. 14th IEEE Conf. Ind. Electron. Appl. ICIEA 2019*, pp. 1926-1931, 2019, doi: 10.1109/ICIEA.2019.8834035.
15. Adelusi, I. Practical approach for elements within incorporated charged zinc particles in an anode zinc reactor of a fabricated zinc bromine battery cell system ($ZnBr_2$) with fitting materials. *Trends Tech. Sci. Res.*, 4(5), 2021.
16. Jiang, H.R., Wu, M.C., Ren, Y.X., Shyy, W., Zhao, T.S. Towards a uniform distribution of zinc in the negative electrode for zinc bromine flow batteries. *Appl Energy*, 213, 366, 2018.
17. Yang, K. *et al.* Stable aqueous Zn–Ag and Zn–polyoxometalate hybrid battery driven by successive Ag+ cation and polyoxoanion redox reactions. *Energy Storage Materials*, 34:203-210, January 2021.
18. Ramesh, M., Muthukrishnan, M., Khan, A. Metal-organic frameworks and permeable natural polymers for reasonable carbon dioxide fixation. In: *Metal-Organic Frameworks for Chemical Reactions*, Elsevier, pp. 417-440, 2020.

19. Ramesh, M., Muthukrishnan, M. Bio-Based magnetic metal organic framework nanocomposites. In: *Metal-Organic Framework Nanocomposites: From Design to Applications*, pp. 167-190, 2020.
20. Xu, Z., Wu, M. Toward dendrite-free deposition in zinc-based flow batteries: status and prospects. *Batteries*, 8(9), pp. 1-23, 2022, doi: 10.3390/batteries8090117.
21. Ramesh, M., Muthukrishnan, M. Polymer electrocatalysis. In: *Methods of Electrocatalysis: Advanced Materials and Allied Applications*, pp. 125-147, 2020.
22. Arfeen, Z.A., *et al.*, Energy storage usages: Engineering reactions, economic-technological values for electric vehicles—A technological outlook. *Int. Trans. Electr. Energy Syst.*, 30(9), pp. 1-30, 2020, doi: 10.1002/2050-7038.12422.
23. Ramesh, M., Kuppuswamy, N., Praveen, S. Metal-organic framework for batteries and supercapacitors. In: *Metal-Organic Frameworks for Chemical Reactions*, pp. 19-36, 2021.
24. Wu, M.C., Zhao, T.S., Jiang, H.R., Zeng, Y.K., Ren, Y.X. High-performance zinc bromine flow battery via improved design of electrolyte and electrode. *J. Power Sources*, 355, 62, 2017.
25. Dixon, D., *et al.*, Tuning the performance of vanadium redox flow batteries by modifying the structural defects of the carbon felt electrode. *Beilstein J. Nanotechnol.*, 10, 1698, 2019.
26. Butler, P., Eidler, P., Grime, P. Zinc/Bromine Batteries. *Adv. Batter. Syst.*, 1, 37.2, 2000.

5

Zinc-Cerium Hybrid Redox Flow Batteries

Raghvendra Mishra and Rajendra K. Singh*

Ionic Liquid and Solid-State Ionics Laboratory, Department of Physics, Institute of Science, Banaras Hindu University, Varanasi, India

Abstract

Due to numerous benefits including energy density, cell potential, and cost-effectiveness, zinc-based hybrid flow batteries (RFBs) are thought to be the most promising systems for energy storage applications. One such device that has been successfully scaled up and commercialized is the Zinc-Cerium (Zn-Ce) redox flow battery. The Zn-Ce flow battery has been investigated widely in the lab and on the market since its inception in 2005. Its open-circuit cell potential may exceed 2.4 V when completely charged, making it the highest of all aquatic RFBs. Decades of research have been devoted to understanding the Ce^{4+}/Ce^{3+} species as a redox mediator and cleaning agent in various types of acids such as sulfuric acid and nitric acid. Zinc-cerium hybrid redox flow batteries are discussed in depth in this chapter, including their history, components, operating principle, and other critical features including cell design and performance.

Keywords: Energy storage technology, Zn-Ce redox flow battery, Ce^{4+}/Ce^{3+} redox couple

5.1 Introduction

Although renewable energy sources are clean, long-lasting, and environmentally beneficial, they cannot be seen as a reliable supply of power because of their intermittent nature [1–3]. But the issue of intermittency of these renewable sources can be handled to a significant extent by the right arrangement of energy storage and conversion technologies including mechanical, thermal chemical, and electrochemical energy storage

**Corresponding author*: rksingh_17@rediffmail.com; rajendrasingh.bhu@gmail.com

Inamuddin and Tariq Altalhi (eds.) Advanced Redox Flow Technology, (107–126) © 2024 Scrivener Publishing LLC

technology [4–6]. Electric and hybrid-electric cars (EVs/HEVs), portable gadgets, and many electronics users have benefited from these technologies, which have been proven to be efficient, adaptable, and ecologically beneficial [7]. Different types of energy storage technologies exist, including chemical (in fuels and explosives), mechanical (flywheels, pumped hydropower), electrical (capacitors and magnetic storage), thermal (heat storage), and electrochemical (batteries, supercapacitors, and fuel cells). In the several existing and developing electrochemical devices such as metal-ion batteries, supercapacitors, fuel cells, etc., the redox flow battery (RFBs) technology seems to be the most realistic option for grid-level energy storage [8]. The energy is often stored as electroactive redox couples in the electrolytes of traditional redox flow batteries, which include all vanadium and bromine-based chemistry, as opposed to hybrid redox flow systems, where at least one electrode is involved in the redox process [9]. In this chapter, the recent technological developments in the field of RFB have been discussed and placed in the context of the field's evolution over the past few decades.

5.1.1 Redox Flow Batteries (RFBs)

Redox flow batteries (RFBs) are effective energy storage devices because, unlike ordinary batteries, they transform chemical energy into electrical energy on demand. RFBs are an alternative solution to the issues of load balancing and uneven power supply, and they also facilitate the use of renewable energy sources. A typical RFB, unlike traditional rechargeable batteries, consists of two independent reservoirs holding catholyte and anolyte that are linked by a flow circulation mechanism (i.e., pump). The electrodes are separated by an ion-selective membrane. Increasing the concentration of the electroactive species and the number of electrolytes increases the capacity of an RFB, making RFBs more important than traditional battery systems [10]. RFBs and fuel cells (FCs) are similar in many ways; however, the electrolyte in an FC system is contained inside the cell stack, while in an RFB system, the catholyte and anolyte independently flow across the cell stack and perform redox reaction. To a large degree, RFB technology allows for the separation of capacity and power. Capacity is dictated by the structural design and the size of the battery pack, while power is specified by the volume of the reservoirs and the quantity and concentration of reactant [9, 10].

5.1.2 The Basic Concept of Redox Flow Batteries

A traditional redox flow cell uses a soluble redox couple in the electrolyte(s) and consists of a positive and negative electrode that are separated by an ion exchange membrane. The separator (ion selective membrane) divides the two half cells where the catholyte and anolytes are circulated through the circulatory system while allowing the transit of ions. Each electrode undergoes electrochemical redox reactions. The membrane permits protons to be transferred to preserve charge and electrolyte balance while limiting reactive species transport amongst the anode and cathode sections. To complete the electrochemical reaction and generate electricity, electrons move through the outer circuit. The following Figure 5.1 shows the typical block diagram for a redox flow battery.

RFBs possess some major advantages such as flexible configuration, easy maneuvering, and prolonged cycle life. Since the energy conversion and storage locations are different in RFBs, their capacity can be significantly enhanced by increasing the reservoir volume without any significant impact on the power of the device [11]. Furthermore, the cycle life of RFBs is not much affected by electrode degradation as they only facilitate a platform for charge transfer, which leads to a higher life in RFBs. The gravimetric energy density (Whkg^{-1}) of the RFB has expressed the equation:

$$u = \int_0^c \frac{V(q)dq}{wt} \tag{5.1}$$

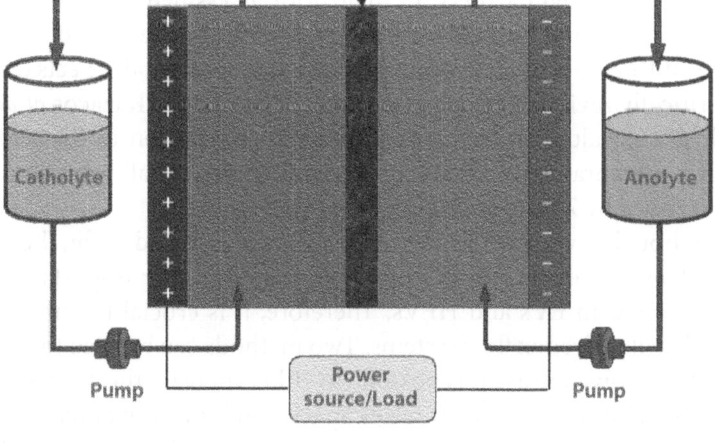

Figure 5.1 Schematic diagram for the redox flow battery.

Where, V stands for the cell potential, q for the charge, and C/wt for cell capacity. The number of redox couples present in the reservoir is proportional to C/wt. Additionally, the Nernst equation may be used to define the equilibrium voltage difference of redox couplings present in the catholyte and anolyte solutions, which could be used to determine the potential (V) of RFB:

$$E = E^{o} - \frac{RT}{nF} ln\, ln\, Q \tag{5.2}$$

Where n is the number of electrons transported, F is the Faraday constant, T is the absolute temperature, E° is the standard potential, and Q is the reaction quotient that relies on the chemical species' activity.

Because the RFBs are instantly recharged by dumping and replenishing the depleted catholyte/analytes solutions with a fully charged one, similar to traditional refueling stations (i.e., petrol, diesel), they may be widely used in fast-charging stations. The power and energy components of an RFB are decoupled, giving the vehicle designer more design options. Additionally, RFBs provide flexibility in adjusting the cell stacking and reservoir size, which may be utilized to control the performance characteristics of redox flow batteries.

5.1.3 Progress and Challenges in the Redox Flow Batteries

RFB research has advanced significantly over the past 40 years, as shown in Figure 5.2, by the surge in publications and the active cooperation of academic and industrial partners. In their prior assessments, Bartolozzi [12] and Ponce de León *et al.* [13] separately covered the basic and historical features of RFBs. Technical, economic, and environmental aspects of RFBs were critically reviewed by Yang *et al.* [14] and Skyllas-Kazacos *et al.* [8], as well as an evaluation of their commercial penetration in the grid. Ion exchange membrane developments and difficulties for all vanadium RFB were reviewed by Zhang *et al.* [15].

Even though several RFB systems have been developed so far, their poor energy density prevents them from being used in other important applications such as in EVs and HEVs. Therefore, it is crucial to enhance the energy density of redox flow systems. Two methods can be used to increase the RFBs' energy density: It is possible to either increase the concentration of redox couples which in turn raise capacity and increase potential of the cell by using the redox couples that have high potential difference [8, 16].

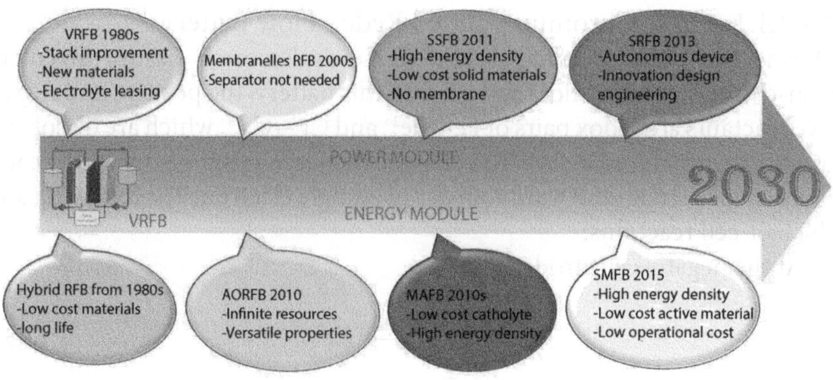

Figure 5.2 Progress in redox flow batteries.

Rapid advancements in new battery chemistries over the past two years have introduced fundamentally new operating principles for RFBs.

5.1.4 Types of Redox Flow Batteries

In general, redox flow batteries may be categorized into the following three groups, which are determined by the phases of the redox active materials that are present in the system.

5.1.4.1 Aqueous Redox Flow Batteries

In aqueous redox flow batteries, chemical energy is mostly stored in the electrolyte. These batteries employ redox couples with reactants/products dissolved in liquid electrolytes. The storage capacity of aqueous redox flow batteries can be enhanced easily by employing much amount of electrolyte. Ion exchange membranes or separators are required in all of these batteries to separate the anode section and cathode section, which also permits ion movement to balance charge neutrality. Following are some aqueous redox flow chemistries as shown in Figure 5.3.

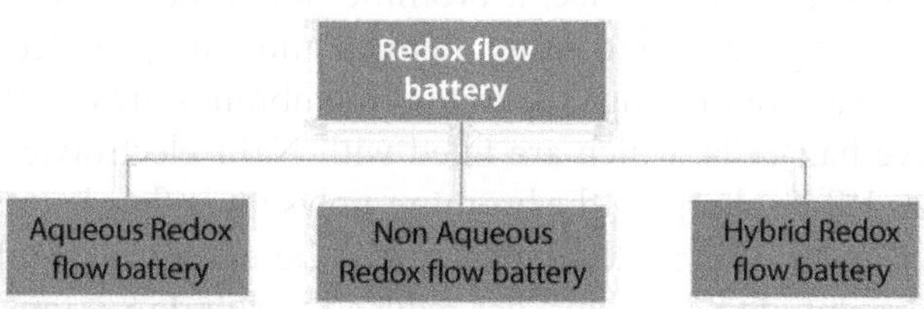

Figure 5.3 Progress in redox flow batteries.

5.1.4.1.1 Iron-Chromium-Based Redox Flow Battery

During the 1970s, Thaller at NASA's Lewis Research Centre invented iron-chromium redox flow batteries. In this battery, the positive and negative reactants are redox pairs of Fe^{3+}/Fe^{2+} and Cr^{3+}/Cr^{2+}, which are dissolved in HCl (hydrochloric acid). The charge/discharge reactions for both reactants entail a straightforward transfer of a single electron. The following are the half-cell reactions:

At the negative electrode,

$$Cr^{3+} + e^- \longleftrightarrow Cr^{2+} \tag{5.3}$$

A positive electrode,

$$Fe^{2+} - e^- \longleftrightarrow Fe^{3+} \tag{5.4}$$

During the charging process, Fe^{2+} undergoes oxidation to become $Fe^{3+,}$ while Cr^{3+} undergoes reduction to become Cr^{2+} at the positive and negative electrodes, respectively. The process is reversed during discharge. The two flowing reactant solutions exist on opposite sides of an ion-exchange membrane, which also avoids the reactive cations from mixing. On most electrodes, the conversion of Cr^{3+} to Cr^{2+} occurs at a sluggish pace; therefore, there is a need for a catalyst on the chromium electrode side. Because the evolution of hydrogen during the reduction of Cr^{3+} to Cr^{2+} tends to result in an intensely competitive reaction, catalysts like bismuth, thallium, lead, and gold is also needed. On the iron electrode, no catalyst is necessary. The iron-chromium RFB suffers from several technical challenges, which have impeded the industry's ability to expand its commercial production. Having an energy density of less than 10 Wh kg^{-1} is the primary drawback of this battery. The cost of maintaining the system is also increased by the need for noble catalysts in the Cr^{3+}/Cr^{2+} electrode reaction.

5.1.4.1.2 Bromine-Polysulfide Flow Battery

Another aqueous-based device, a bromine-polysulfide flow battery, was developed by Regenesys, Ltd., after Remick and Ang patented it in 1983 [9, 13]. The Na^+-conducting separator (membrane) divides the positive and negative half-cells, which are filled with NaBr electrolyte and Na_2S_4, respectively. While charging the bromine polysulfide flow battery, Br^- ions get oxidized to Br_2 at the cathode terminal, whereas, at the anode side, the soluble polysulphide S_4^{2-} anions, get reduced to sulfide S_2^{2-} ions. The electrode reactions involved in bromine-polysulfide flow battery are as follows.

At the negative electrode,

$$S_4^{2-} + 2e^- \leftrightarrow 2S_2^{2-} \tag{5.5}$$

At the positive electrode,

$$3Br^- - 2e^- \leftrightarrow Br_3^- \tag{5.6}$$

The cheap cost and superior solubility of the electrolyte components are just two of the many benefits that come along with using RFBs of this particular type. The OCV and energy density of the bromine-polysulfide flow battery are roughly 1.50 V and 20 to 30 WhL^{-1}, respectively [17, 18]. Reduced cycle capacities are often the consequence of cross-contamination and the precipitation of sulfur species. Due to the discharge of poisonous gases including Br_2 and H_2S, safety issues continue to be problematic.

5.1.4.1.3 Vanadium-Based Redox Flow Battery

Currently, the vanadium-based RFB system, which was designed by Skyllas-Kazacos *et al.* in the 1980s, is the RFB system that has been the most successful in terms of commercialization [8]. To prevent cross-contamination, the system uses the same component in both half-cells, namely $V^{2+}/^{3+}$ (at negative) and $V^{4+/5+}$ (at positive) pairs. The following are the electrode reactions in VRFB:

At the negative electrode,

$$V^{3+} + e- \longleftrightarrow V^{2+} \tag{5.7}$$

At the positive electrode,

$$VO^{2+} + H_2O \leftrightarrow VO_2^+ + 2H^+ + e^- \tag{5.8}$$

When fully charged, the VRDF displays an open-circuit potential (OCP) of 1.60 volts, whereas, at half-charged condition, it displays a value of 1.35 V. In 1991, Skyllas Kazacos *et al.* established a stack of 1 kW with 90% energy efficiency and a power density of 0.13 Wcm^{-2} [19, 20]. The system is reasonably inexpensive because the electrolytes can be reused. Another benefit is the system's safety, which prevents deep discharge and overcharging from irreversibly harming the system (i.e., cells or electrolytes). The major shortcoming of the vanadium-based RFB is its poor specific energy (25-30 $Whkg^{-1}$), which results due to the solubility of the redox active elements at a low span of working temperatures [21, 22].

5.1.4.1.4 Quinone-Bromine Flow Battery

The quinone-bromine flow battery, which was developed by Huskinson *et al.*, is one example of a recent advancement made to aqueous flow batteries [22, 23]. It has a relatively high energy density and low open-circuit voltage (0.92 V at 90% state of charge) when compared to the all-vanadium RFB (50 WhL^{-1}). Maximum power density approaches 0.6 Wcm^{-2} at a distance of 1.3 Acm^{-2}. However, results for long-term stability are still pending publication, even though cycling measurements show a coulombic efficiency of over 99% for the first ten cycles. One benefit of this setup includes the ease and cost-effectiveness of the quinone/hydroquinone and Br_2/Br pairs containing materials. Additionally, the fast kinetics of quinones can be achieved with just carbon electrodes, suggesting that a costly metal catalyst is unnecessary. Ultimately, this helps bring down the overall price of the system. Poisonous Br_2 is always a potential threat.

$$AQDSH_2 \longleftrightarrow AQDS^- + 2H^+ + 2e^- \quad (5.9)$$

$$Br_2 + 2H + 2e^- \longleftrightarrow 2HBr \quad (5.10)$$

Where, [$AQDSH_2$] = reduced form of AQDS, and AQDS = 9,10-anthraquinone-2,7-sulphonic acid.

5.1.4.2 Nonaqueous Redox Flow Batteries

To put it another way, non-aqueous redox flow batteries are those in which the electrode stores chemical energy in the form of active material. The charging process of these batteries involves solid species that participate in redox couples. Discharge is the process whereby energy is released via electrodeposits at the positive and negative electrodes. In recent years, several different flow battery systems have been introduced [24–26]. In the majority of such systems a single electrolyte is required to be shuttled around the cell, eliminating the need for a separator (ion-exchange membrane). Due to the lack of a membrane separator, this system is not susceptible to the outflow and cross-contagion of the redox couples that have plagued other flow battery systems. Additionally, this significantly lessens the price and complication of the architecture of this battery system.

5.1.4.2.1 Soluble Lead-Acid Flow Battery

Pletcher *et al.* (2004) proposed a soluble lead-acid flow battery without a membrane [25]. This battery uses Pb^{2+}/PbO_2 and Pb/Pb^{2+} active species at the cathode and anode, respectively, to facilitate the electrochemical reactions of Pb(II) in methanesulfonic acid. Here, it is worth noting that, the electrode reactions happening in the cell do not lead to sulfate inside the paste, which is a key difference from the chemistry of a standard lead-acid battery. The open-circuit cell voltage of the laboratory flow battery is 1.69 V, and the charge efficiency is 85%. The energy efficiency is 65% [25, 27, 28]. During the process of development, issues with Pb dendrite growth and PbO_2 shedding caused a substantial loss in efficiency and, in most cases, battery failure. Extensive testing has revealed that hexadecyl-trimethylammonium cation and similar additives can inhibit the growth of Pb dendrites on the anode terminal [29, 30]. The phase structure of PbO_2 was also found to have major effects on cell performance. Moreover, several alternative flow batteries were proposed based on the membrane-free concept [26, 31]. For instance, one single-flow Cu-H_2SO_4-PbO_2 battery was reported to have a round-trip DC energy efficiency of up to 83% and an open-circuit cell voltage of 1.35 V.

5.1.4.2.2 Zinc-Nickel

Zinc-nickel single electrolyte flow batteries were first stated in 2007 [32, 33]. As the supporting electrolyte, a zincate solution in potassium hydroxide was used. Zinc is produced by the electrodeposition of zincate ions, while nickel is produced through the oxidation of nickel$(OH)_2$ to nickel(OOH) at the anode and cathode, respectively, during charging, the processes are inverted during discharging. Although it seems cost-effective and efficient in performance, this system has the same limitation as the lead-lead dioxide system in that its capacity is determined by the dimensions of its electrodes.

5.1.4.3 Hybrid Redox Flow Batteries

Hybrid flow batteries are redox flow batteries (RFBs) featuring half-cell reactions involving the deposition of solid or gaseous species. Unlike conventional redox flow batteries, hybrid RFBs use a combination of both solid and dissolved redox species. The solid redox species that are frequently used in traditional systems, for instance, have taken the place of the redox mediators that were dissolved in catholyte or anolyte. Due to their significantly larger energy density over liquid electrolytes and comparatively easy chemistry, metals are generally the most preferred option in this regard. The following hybrid RFB chemistries are among the many being researched.

5.1.4.3.1 Alkaline-Ion Cathode Flow Battery

Alkaline metals are primarily used as the anode in alkaline-ion batteries, along with some redox species like ferricyanide that are dissolved in the catholyte. A hybrid redox flow battery made of ferricyanide and lithium metal was described by Lu and Goodenough [34]. A ceramic glass membrane served as a barrier between the electrolyte (organic solution) used at the negative electrode and the aqueous solution used at the positive electrode. The maximum reported power density for this battery, which operates at a voltage of about 3.40 V with 97% coulombic efficiency after 1000 cycles, is 17 $mWcm^{-2}$ at 10 $mAcm^{-2}$. An identical tactic was used by Wang *et al.*, who tried lithium metal as the anode in a nonaqueous electrolyte and the $FeCl_3/FeCl_2$ couple in the aqueous catholyte where HCl was used to stop the hydrolysis of Fe^{3+} [35, 36]. They also advised replenishing the cathodic tank regularly using the oxidizing agent $NH_4S_2O_8$. Similar to this, research has also been done on other chemistries including lithium-iodine flow batteries, lithium polysulfide batteries, and other flow batteries with solid/semisolid active materials.

5.1.4.3.2 Zn-Based Hybrid Redox Flow Batteries

Zinc is a common choice for the standard negative electrode in cells owing to its high negative electrode potential in basic (-1.22 V vs. SHE) as well as acidic (-0.76 V vs. SHE) medium. Zinc is also a cheap and abundant metal. The zinc-bromine battery is the most common type of hybrid flow battery, which can store a theoretical specific energy of 440 $Whkg^{-1}$ and OCV of 1.70 V with 80% energy efficiency [8, 37]. The aqueous phase in the storage tank is separated from the emulsion because the emulsion cannot dissolve in water. Even though only a few businesses have created extensive zinc-bromine RFB systems, many problems still need to be fixed, including poor efficiency, high self-discharge rate, and brief life cycle. The large surface area of the positive carbon electrode compensates for sluggish bromine/bromide conversion kinetics. However, the carbon electrodes' surface area will gradually decrease as a result of bromine oxidation.

Anodic zinc electrodes, on the other hand, have a problem with the formation of dendritic zinc. The zinc-chlorine battery discharges at a voltage of 1.80 V with 84% coulombic and 66% energy efficiencies and has also been the subject of reports [32]. The system's primary weakness is the major ecological risks posed by the development of dangerous Cl_2 gas [32]. The zinc-based HFBs also come in Zn/Ce, Zn/Ni, and Zn/MnO2 chemistries in addition to these. Zinc-cerium hybrid redox flow batteries are mainly intriguing among them owing to their many benefits, including high cell voltage and minimal dendrite formation [38].

5.2 Zinc-Cerium Hybrid Redox Flow Battery

The higher positive as well as negative standard potential of the Ce^{4+}/Ce^{3+} couple and zinc deposition, respectively, and the fast solubility of metal ionic species in methanesulfonic acid (aqueous) have inspired the idea of zinc-cerium chemistry [39]. In a variety of media, the Ce^{4+}/Ce^{3+} combination has also been found to have relatively quick kinetics [40]. In 2004, Plurion Inc. (UK) debuted the Zinc-cerium RFB, which had previously been patented by Clarke *et al.* [41]. This system is claimed to have a higher OCV of about 2.4 V and the ability of high rate discharge. Instead of sulphuric acid, methanesulfonic acid is usually used as an electrolyte due to its less corrosive nature, and also it has the quality to permit higher concentrations of Ce-ions. Additionally, it also slows down the expansion of zinc dendritic tissue, which shortens the zinc-halogen systems' cycle life.

5.2.1 Working Principle of Zn-Ce Redox Flow Cell

In the Figure 5.4, a typical Zn-Ce cell with a cation-exchange membrane containing Zn^{2+} and Ce-electrolytes is shown in operation. In practice, the battery is typically constructed from several unit cells stacked on a dual side stacking. The dimensions of the electrodes and the number of cells, respectively, govern the highest current and potential of the cell. The mixing of the Zn^{2+} and Ce^{3+} ions over time indicates that the membrane is not entirely proton transport selective. The battery has thus been worked in absence of a membrane that contains both cerium (Ce^{3+}) and zinc (Zn^{2+}) ions. The electrochemical reactions that occur in Zn-Ce redox flow cells are as follows:

At the negative electrode,

$$Zn^{2+} + 2e^- \longleftrightarrow Zn \qquad E^o = -0.76\ V\ Vs\ SHE \tag{5.11}$$

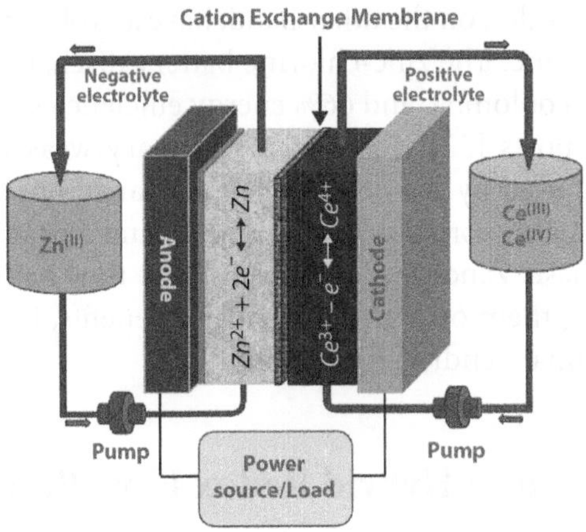

Figure 5.4 Working of Zn-Ce redox flow battery.

At the positive electrode,

$$Ce^{3+} - e^{-} \longleftrightarrow Ce^{4+} \qquad E^{o} = 1.44\ V\ Vs\ SHE \qquad (5.12)$$

Overall, Cell reaction,

$$2Ce^{3+} + Zn^{2+} \leftrightarrow 2Ce^{4+} + Zn \qquad E^{o}_{cell} = 2.2\ V \qquad (5.13)$$

Additionally, the side reactions that occur during the charge can lead to the evolution of hydrogen at the anode and oxygen at the cathode, which reduces the efficiency of the charge and changes the pH in the area around the electrodes. Additionally, if the potential is not appropriately negative, the corrosion of recently deposited zinc may result in electrode dissolution.

$$Zn + 2H^{+} = Zn^{2+} + H_2 \qquad (5.14)$$

5.2.1.1 *Components of Zn-Ce Redox Flow Battery*

The detailed discussion on the various components of the Zn-Ce redox flow battery is as follows below.

5.2.1.1.1 The Positive Electrode (Cerium)

Cerium acts as the positive electrode in Zn-Ce redox flow batteries. Since oxygen evolution during battery charging affects the Ce^{4+}/Ce^{3+} redox process, the voltage and charge efficiency are reduced. The solubility of the Ce-species is inadequate for the functioning of flow batteries in a realistic context, even though the voltammetry of the Ce^{4+}/Ce^{3+} species has been investigated in numerous aqueous as well as acidic solutions. Despite this, it was shown that the couple typically has fairly quick kinetics. In methanesulfonic acid, Ce^{4+}/Ce^{3+} is a quasi-reversible couple, according to preliminary voltammetry studies [42–44]. When methanesulfonic acid concentration was measured as a function of the solubility of cerium species, Spotnitz *et al.* [44] discovered that it rapidly decreased over 4 mold m^{-3} acid and the cell performance decreases with anodes in the order PbO_2>Pt>coated Ti>C. At a current density of 100 $mAcm^{-2}$, the efficiency of the Pt electrode was ~96%. This development was expanded upon for larger flow cells by Harrison and Théorêt [45], who also observed the properties of anode material, current density and flow rate, etc. Xie *et al.* [46] investigated CV measurements to identify the ideal circumstances for using the Ce^{4+}/Ce^{3+} couple in a flow battery. They observed that the Ce^{4+}/Ce^{3+} species had a stronger positive standard potential in methanesulfonic acid compared to sulfuric acid. Both Ce^{4+} and Ce^{3+} are entirely stable up to 60°C, which is a definite advantage for the couple. However, at higher temperatures, Ce^{4+} may progressively precipitate out of electrolytes containing concentrated Ce^{3+}. As a half-cell electrolyte for flow batteries, cerium in methanesulfonic acid is preferable since it can be readily manufactured at concentrations as high as 1 $moldm^{-3}$, in contrast to less than 0.5 $moldm^{-3}$ in sulfuric acid. To optimize the resultant cerium half-cell as the positive electrode reaction in a redox flow battery, a deeper knowledge of the electrochemical kinetics of the Ce^{4+}/Ce^{3+} redox couple in methanesulfonic acid under different circumstances is required.

5.2.1.1.2 Negative Electrode

During the charging process, the zinc is deposited on a collector, just like in other hybrid zinc redox flow batteries [11, 33, 47]. One of the Zn-based RFB systems that have received the most research is the zinc-halogen hybrid. Electrode substrate, cell shape, charging technique, composition, and the morphology of the zinc electrode are the variables that have been found to have an impact on zinc deposition. Guillaume *et al.* investigated that increasing the zinc species concentration (>2 mold m^{-3}) would result in higher efficiencies of around 80% [48]. On the other hand, some issues with the zinc-bromine RFB include the HER at the anode during the charging process, which causes low efficiencies as well as uneven zinc deposition during random cycling [49]. Van Parys *et al.* [40] looked into the mechanism of zinc deposition and hydrogen evolution. The induced micro stirring was found to increase the current density if gas bubbles were created in the mass-transport-regulated zone, but to decrease it when the current was kinetically controlled due to the enhanced value of resistance of the electrolyte. On cycling, the zinc electrode active material was found to be redistributed in the zinc-nickel cell, the separator was found to be penetrated by zinc dendrites, and the zinc electrode was found to be de-zincified. Ito *et al.* [50] investigated that for this system, high electrolyte flow rates, defined as those higher than 15 cm s^{-1}, increased the battery's cycle life at a 100% depth of discharge.

5.2.1.1.3 Electrolyte

The capacity of a redox flow battery (RFB) is proportional to the number of its redox active couples. In the Zn-Ce system, the acid content also affects the solubility of the cerium species and even the device's overall resilience. Methanesulfonic acid (MSA) has been utilized as a supporting electrolyte in the majority of Zn-Ce cells due to its ability to dissolve cerium ions at reasonably high density [44]. Each side of the divided cell uses a separate electrolyte. Leung *et al.* made their positive half-cell out of 0.8 moldm^{-3} Ce^{3+} dissolved in 4 moldm^{-3} MSA, and their negative half-cell out of 1.5 moldm^{-3} Zn^{2+} dissolved in 1moldm^{-3} MSA [13, 32]. Because of the different electrolyte composition, ions might go across the separator membrane into the negative half-cell, altering the working of RFB and shortening its useful life. Nikiforidis *et al.* added 0.8 mold m^{-3} of Zn^{2+} to the positive half-cell electrolyte to limit the movement of Zn-ions from the Zn-electrolyte and utilized a to some extent lesser concentration of cerium [51]. A highly concentrated solution of 2.5 moldm^{-3} Zn^{2+} in 1.5 moldm^{-3} MSA served as the electrolyte in the negative half-cell. By using only one electrolyte that contains both redox species, undivided cells might address some of the problems that split cells experience. Using 0.2 moldm^{-3} Ce^{3+}, 1.5 moldm^{-3} Zn^{2+}, and 0.5 moldm^{-3} MSA, Leung *et al.* created a single electrolyte [47]. In this instance, the low acid content reduced the HER and made it easier for zinc to electrodeposit. The cerium concentration was also maintained low since, in that case, Ce^{4+} ions would compete for space at the electrode surface, dramatically raising the zinc electrodeposition overpotential. Nevertheless, the cell's low cerium content resulted in a low energy density. In addressing some of the drawbacks of the existing Zn-Ce RFBs [52], it has not yet been determined how sulfuric acid affects the stability of the carbonaceous electrodes in maximum sophisticated cells.

5.2.2 Factors Affecting the Performance of Zn-Ce Redox Flow Battery

5.2.2.1 Temperature

At relatively high temperatures, the cycling performance of divided Zn-Ce RFBs improves. Cell potential and charge efficiencies increased with temperature when cells were tested in the 25 to 60 °C temperature range [32, 47]. The energy efficiency would have the same result. The operating temperature of 50 °C was determined to be the most suitable after considering various other factors. At the lowest temperature, the coulombic efficiency significantly decreased, and for each of the two negative electrode materials, polyvinylidene difluoride composites, 55 and 45 °C, respectively, were found to be the optimal temperature. Faster kinetics and increased diffusion coefficient brought on by a decrease in electrolyte viscosity can be used to explain these effects [47]. This relationship between efficiency and temperature was quite different for an undivided cell with charge efficiencies declining as the temperature rose [47, 49]. One benefit of the undivided cell may be that it performs best at room temperature, which was discovered to be the case for the undivided system.

5.2.2.2 *Electrolyte Flow Rate*

In their divided cell, Leung *et al.* examined the linear velocity of electrolytes [32]. When used as positive electrodes, the ideal flow velocity for maximum discharge voltage was 3.9 cms^{-1} for carbon felt and 2.6 cms^{-1} for platinized titanium. At 7.8 cms^{-1}, the highest energy efficiency was attained. Nikiforidis *et al.* used electrolyte flow rates in the divided cell between 7.5 and 13.5 cms^{-1} [47, 53]. Coulombic and voltage efficiencies under these laminar flow conditions showed little deviation from 90% and 60%, respectively. Therefore, 10.5 cms^{-1} flow velocity was selected for the following charge/discharge experiments. With a mixed electrolyte, Xie *et al.* used a divided cell with a flow velocity of 11.5 cms^{-1} [54]. In the undivided cell, the common electrolyte was pumped at various flow rates, ranging from 0.64 to 7.0 cms^{-1}. The carbon felt's huge surface area prevents further velocity increases of more than 2.0 cm/s^{-1}, only resulting in negligible improvements in the discharge potential. It was decided that a flow velocity of 3.9 cms^{-1} would be ideal because it had a slightly higher charge efficiency.

5.2.2.3 *Current Density*

The charge-discharge performance in a split cell was tested using a platinized titanium-positive electrode. The constant current density was kept the same throughout the study [32]. The cell displayed the greatest possible mass transfer and the least amount of interference from secondary processes. Its coulombic efficiency was measured at 50 $mAcm^{-2}$. The University of Strathclyde cell was tested in all of the experiments with a charging current of 10 $mAcm^{-2}$ [47, 53].

5.2.2.4 *Charge Conditions and Cycle Life*

Both the charge time and the efficiency of the Zn-Ce RFBs have been shown to have a significant connection. Charging and discharging cycles of 0.25 hours in length worked well up to 57 times, but their voltage efficiency was only 46.2% and their coulombic efficiency was only 85% [32]. Even though this reduced the total number of cycles to 25, it was discovered that the most favorable circumstances for coulombic, voltage, and energy efficiency (99, 69, and 60%, respectively) were obtained when a three-hour charge session was performed before the 0.25-hour cycles. The first few cycles of the cerium reaction were ineffective, which limited the battery's ability to discharge. The undivided cell was charged and discharged over a half hour by the same research team [55].

5.3 Summary

This chapter provides a comprehensive overview of zinc-cerium hybrid redox flow batteries, including topics such as their development, components, working principle, cell design, and performance. The conventional approach to the chemistry of Zn-Ce batteries requires an aqueous methanesulfonic acid electrolyte to be enclosed inside a cell and separated from one another by a proton-exchange membrane. The Zn-Ce cell is attractive because it has a high open-circuit cell potential (about 2.2 V), in addition to a relatively good volumetric energy density. It has been shown that improving the kinetics of the cerium half-cell and cell capacity may be accomplished by the use of a mixed-acid electrolyte, the addition of additions, and the utilization of a single acid medium in which the cerium salt is highly soluble. Since Ce^{3+} is oxidized to Ce^{4+} at very high potentials, oxygen evolution is a significant side reaction that takes place during charging (1.2-1.7 V vs. SHE). The presence of Ce^{4+} during charge poses limits on the positive electrode material selection owing to the need for a catalytic material for the Ce^{3+}/Ce^{4+} redox reaction and good chemical stability in strong acid electrolytes.

Acknowledgment

One of the authors (R.K. Singh) gratefully acknowledges the financial support provided by SERB, New Delhi, under IMPRINT (II)C to carry out the research work. R. Mishra sincerely acknowledges the financial assistance from SERB in the form of SRF under IMPRINT (II)C.

References

1. Patel, A., Gupta, H., Singh, S.K., Srivastava, N., Mishra, R., Meghnani, D., Tiwari, R.K., Tiwari, A., Tiwari, V.K., Singh, R.K., Superior cycling stability of saturated graphitic carbon nitride in hydrogel reduced graphene oxide anode for Sodium-ion battery. *FlatChem.*, 33, 100351, 2022.
2. Mishra, R., Singh, S.K., Gupta, H., Srivastava, N., Meghnani, D., Tiwari, R.K., Patel, A., Tiwari, A., Tiwari, V.K., Singh, R.K., Surface modification of nano $Na[Ni_{0.60}Mn_{0.35}Co_{0.05}]O_2$ cathode material by dextran functionalized RGO via hydrothermal treatment for high performance sodium batteries. *Appl. Surf. Sci.*, 535, 147695, 2021.
3. Gupta, H., Singh, S.K., Singh, V.K., Tripathi, V.K., Srivastava, N., Tiwari, R.K., Mishra, R., Meghnani, D., Singh, R.K., Development of polymer electrolyte and cathode material for Li-batteries. *J. Electrochem. Soc.*, 166, A5187–A5192, 2018.
4. Singh, S.K., Shalu, L., Gupta, H., Singh, V.K., Tripathi, A.K., Verma, Y.L., Singh, R.K., Improved electrochemical performance of EMIMFSI ionic liquid based gel polymer electrolyte with temperature for rechargeable lithium battery. *Energy*, 150, 890–900, 2018.
5. Singh, S.K., Dutta, D., Singh, R.K., Enhanced structural and cycling stability of Li_2CuO_2-coated $LiNi_{0.33}Mn_{0.33}Co_{0.33}O_2$ cathode with flexible ionic liquid-based gel polymer electrolyte for lithium polymer batteries. *Electrochim. Acta.*, 343, 136122, 2020.
6. Meghnani, D., Gupta, H., Singh, S.K., Srivastava, N., Mishra, R., Fabrication and electrochemical characterization of lithium metal battery using IL-based polymer electrolyte and Ni-rich NCA cathode. *Ionics*, 26, 4835-4851, 2020.
7. Julien, C., Introduction energy storage : Materials and lithium-ion batteries, in: *Materials Science for Energy Storage*, pp. 1-8, 2017.
8. Kazacos, M., Chakrabarti, M.H., Hajimolana, S.A., Mjalli, F.S., Saleem, M., Progress in flow battery research and development. *J. Electrochem. Soc.*, 158, R55, 2011.
9. Emmett, R.K., Roberts, M.E., Recent developments in alternative aqueous redox flow batteries for grid-scale energy storage. *J. Power Sources*, 506, 230087, 2021.
10. Huang, Q., Wang, Q., Next-generation, high-energy-density redox flow batteries. *ChemPlusChem.*, 80, 1–12, 2014.
11. Weber, A.Z., Mench, M.M., Meyers, J.P., Ross, P.N., Gostick, J.T., Liu, Q., Redox flow batteries: A review. *J. Appl. Electrochem.*, 41, 1137–1164, 2011.
12. Bartolozzi, M., Development of redox flow batteries. A historical bibliography. *J. Power Sources*, 27, 219–234, 1989.
13. de León, C.P., Ferrer, A.F., García, J.G., Szánto, D.A., Walsh, F.C., Redox flow cells for energy conversion. *J. Power Sources*, 160, 716–732, 2006.
14. Yang, Z., Electrochemical energy storage for green grid: Status and challenges. *Chem. Rev.*, 111, 3577-3613, 2011. https://doi.org/10.1149/ma2011-02/4/155.

15. Li, X., Zhang, H., Mai, Z., Zhang, H., Vankelecom, I., Ion exchange membranes for vanadium redox flow battery (VRB) applications. *Energy Environ. Sci.*, 4, 1147–1160, 2011.
16. Serhan, M., Sprowls, M., Jackemeyer, D., Long, M., Perez, I.D., Maret, W., Tao, N., Forzani, E., Total iron measurement in human serum with a smartphone, in *Annu. Meet. Conf. Proc.*, 2019.
17. Zhao, P., Zhang, H., Zhou, H., Yi, B., Nickel foam and carbon felt applications for sodium polysulfide / bromine redox flow battery electrodes. *Electrochim. Acta.*, 51, 1091–1098, 2005.
18. Gu, S., Gong, K., Yan, E.Z., Yan, Y., A multiple ion-exchange membrane design for redox flow batteries. *Energy Environ. Sci.*, 7, 2986–2998, 2014.
19. Kazacos, M.S., Kasherman, D., Hong, D.R., Kazacos, M., Characteristics and performance of 1 kW UNSW vanadium redox battery. *J. Power Sources*, 35, 399–404, 1999.
20. Kim, S., Vijayakumar, M., Wang, W., Zhang, J., Chen, B., Nie, Z., Chen, F., Hu, J., Li, L., Yang, Z., Chloride supporting electrolytes for all-vanadium redox flow batteries. *PhysChemChemPhys.*, 13, 18186–18193, 2011.
21. Kazacos, M., Menictas, C., Kazacos, M., Thermal stability of concentrated V(V) electrolytes in the vanadium redox cell. *J. Electrochem. Soc.*, 143, L86–L88, 1996.
22. Gerhardt, M.R., Galvin, C.J., Huskinson, B., Marshak, M.P., Suh, C., Chen, X., A metal-free organic–inorganic aqueous flow battery. *Nature*, 505, 195-198, 2014.
23. Chen, Q., Gerhardt, M.R., Hartle, L., Aziz, M.J., A quinone-bromide flow battery with 1 W/cm^2 power density. *J. Electrochem. Soc.*, 163, 5010–5013, 2015.
24. Wang, Z., Tam, L.Y.S., Lu, Y.C., Flexible solid flow electrodes for high-energy scalable energy storage. *Joule*, 3, 1677–1688, 2019.
25. Pletcher, D., Wills, R., A novel flow battery: A lead acid battery based on an electrolyte with soluble lead. *PhysChemChemPhys.*, 6, 1779–1785, 2004.
26. Pan, J., Sun, Y., Cheng, J., Wen, Y., Yang, Y., Wan, P., Study on a new single flow acid Cu-PbO_2 battery. *Electrochem. Commun.*, 10, 1226–1229, 2008.
27. Pletcher, D., Wills, R., A novel flow battery - A lead acid battery based on an electrolyte with soluble lead(II): III. The influence of conditions on battery performance. *J. Power Sources*, 149, 96–102, 2005.
28. Hazza, A., Pletcher, D., Wills, R., A novel flow battery - A lead acid battery based on an electrolyte with soluble lead(II): IV. The influence of additives, *J. Power Sources*, 149, 103–111, 2005.
29. Ravikumar, M.K., Rathod, S., Jaiswal, N., Patil, S., Shukla, A., The renaissance in redox flow batteries. *J. Solid State Electrochem.*, 21, 2467–2488, 2017.
30. Ruslim, M.M., Hamzah, A., Nizam, A.S.M., State of the art of all-Vanadium redox flow battery: A Research Opportunity, in *Int. Conf. Green Technol. Ecosyst. Glob. Sustain. Dev.*, 2012, pp. 1–8.
31. Pan, J., Ji, L., Sun, Y., Wan, P., Cheng, J., Yang, Y., Fan, M., Preliminary study of alkaline single flowing Zn-O_2 battery. *Electrochem. Commun.*, 11, 2191–2194, 2009.

32. Khor, A., Leung, P., Mohamed, M.R., Flox, C., Xu, Q., An, L., Wills, R.G.A., Morante, J.R., Shah, A.A., Review of zinc-based hybrid flow batteries : From fundamentals to applications. *Mater. Today Energy*, 8, 80–108, 2018.
33. Arenas, L.F., Loh, A., Trudgeon, D.P., Li, X., de León, C.P., Walsh, F.C., The characteristics and performance of hybrid redox flow batteries with zinc negative electrodes for energy storage. *Renew. Sustain. Energy Rev.*, 90, 992–1016, 2018.
34. Chen, H., Zou, Q., Liang, Z., Liu, H., Li, Q., Lu, Y.C., Sulphur-impregnated flow cathode to enable high-energy-density lithium flow batteries. *Nat. Commun.*, 6, 2–10, 2015.
35. Wang, Y., Wang, Y., Zhou, H., A Li-liquid cathode battery based on a hybrid electrolyte. *ChemSusChem.*, 4, 1087–1090, 2011.
36. Wang, Y., He, P., Zhou, H., Li-redox flow batteries based on hybrid electrolytes: At the cross road between Li-ion and redox flow batteries. *Adv. Energy Mater.*, 2, 770–779, 2012.
37. Lim, H.S., Lackner, A.M., Knechtli, R.C., Zinc-bromine secondary battery. *J. Electrochem. Soc.*, 124, 1154–1157, 1977.
38. Chen, L., Guo, Z., Xia, Y., Wang, Y., High-voltage aqueous battery approaching 3V using an acidic–alkaline double electrolyte. *Chem. Commun.*, 49, 2204–2206, 2013.
39. Mezei, F., Basics Concepts, 2011. In *Neutrons in Soft Matter*, pp. 1 -28.
40. Parys, H.V., Telias, G., Nedashkivskyi, V., Mollay, B., Vandendael, I., Damme, S.V., Deconinck, J., Hubin, A., On the modeling of electrochemical systems with simultaneous gas evolution. Case study: The zinc deposition mechanism. *Electrochim. Acta.*, 55, 5709–5718, 2010.
41. Clarke, S.M.R.L., Dougherty, B.J., Harrison, S., Millington, J.P., US Patent Appication, 2004.
42. Vijayabarathi, T., Velayutham, D., Noel, M., Influence of aromatic reactants and products involved in the two stage electro-chemical oxidation on the voltammetric behaviour of Ce(IV)/Ce(III) redox couple. *J. Appl. Electrochem.*, 31, 979–986, 2001.
43. Devadoss, V., Noel, M., Jayaraman, K., Basha, C.A., Electrochemical behaviour of Mn3+/Mn2+, Co3+/Co2+ and Ce4+/Ce4+ redox mediators in Methanesulfonic acid. *J. Appl. Electrochem.*, 33, 319–323, 2003.
44. Spotnitz, R.M., Kreh, R.P., Lundquist, J.T., Press, P.J., Mediated electrosynthesis with cerium (IV) in methanesulphonic acid. *J. Appl. Electrochem.*, 20, 209–215, 1990.
45. Harrison, S., and Théorêt, A., The electrosynthesis of naphthoquinone and tetrahydroanthraquinone, *J. New Mater. Electrochem. Syst.* 2 (1999) 1–9.
46. Xie, Z., Zhou, D., Xiong, F., Zhang, S., Huang, K., Cerium-zinc redox flow battery: Positive half-cell electrolyte studies. *J. Rare Earths*, 29, 567–573, 2011.
47. Walsh, F.C., Løon, P., Berlouis, L., Nikiforidis, G., The development of Zn–Ce hybrid redox flow batteries for energy storage and their continuing challenges. *ChemPlusChem.*, 80, 1–25, 2014. https://doi.org/10.1002/cplu.201402103.

48. Guillaume, P., Leclerc, N., Boulanger, C., Lecuire, J.M., Lapicque, F., Investigation of optimal conditions for zinc electrowinning from aqueous sulfuric acid electrolytes. *J. Appl. Electrochem.*, 37, 1237–1243, 2007.
49. Leung, P., Li, X., De Leo, P., Berlouis, L., John, C.T., Walsh, F.C., Progress in redox flow batteries, remaining challenges and their applications in energy storage. *RSC Adv.*, 2, 10125–10156, 2012.
50. Ito, Y., Desai, D., Wei, X., Steingart, D., Banerjee, S., Zinc morphology in zinc nickel-oxide batteries. ECS Meet. Abstr., MA2011-02, 700, 2011.
51. Nikiforidis, G., Investigations of the Zinc-Cerium Hybrid Flow Battery, PhD thesis, University of Strathclyde, Glasgow, Scotland, 2012.
52. Shahbaz, K., Mjalli, F.S., Hashim, M.A., AlNashef, I.M., Using deep eutectic solvents based on Methyl Triphenyl Phosphonium Bromide for the removal of glycerol from palm-oil-based biodiesel. *Energy Fuels*, 25, 2671–2678, 2011.
53. Nikiforidis, G., Berlouis, L., Hall, D., Hodgson, D., Impact of electrolyte composition on the performance of the zinc-cerium redox flow battery system. *J. Power Sources*, 243, 691–698, 2003.
54. Govindan, M., He, K., Moon, I.S., Evaluation of dual electrochemical cell design for cerium-vanadium redox flow battery to use different combination of electrodes. *Int. J. Electrochem. Sci.*, 8, 10265–10279, 2013.
55. Leung, P.K., De León, C., Walsh, F.C., An undivided zinc–cerium redox flow battery operating at room temperature (295 K). *Electrochem. Commun.*, 13, 770–773, 2011.

6

Vanadium Redox Flow Batteries (VRFB)

Abrar Hussain[1], Muhammad Tahir Khan[1*], Samad Yaseen[2], Ata-ur-Rehman[2†] and Syed Mustansar Abbas[3‡]

[1]Department of Physics, Riphah International University, Islamabad, Pakistan
[2]Department of Chemistry, Govt. Graduate College, Asghar Mall, Rawalpindi, Pakistan
[3]Nanoscience & Technology Department, National Centre for Physics, Islamabad, Pakistan

Abstract
The world meets energy demands by using fossil fuels and renewable energy resources. The present need of the global world is to work on energy sources that are economical and efficient. The world has now reduced the use of fossil fuels and is shifting toward renewable energy generation. Among these sources, the vanadium redox flow battery (VRFB) technology that has been developed recently is considered a better candidate for efficient storage of energy. The potential application of VRFB in energy storage is due to a change in its oxidation state from bivalent up to pentavalent (V^{2+}, V^{3+}, V^{4+}, V^{5+}). This chapter covers the working principle, the main components of the VRFB system, and a comparison of VRFB with other electrochemical energy storage technologies. The recent developments on VRFB have been briefly explained. Finally, the most crucial learning research fields are discussed, as well as future development suggestions.

Keywords: Vanadium redox flow battery, oxidation state, electrolyte, carbon-based materials, state of charge, cation exchange membrane, bipolar plates, chemical vapor deposition technique

**Corresponding author*: mtahir.khan@riphah.edu.pk
†*Corresponding author*: ata.qau@gmail.com
‡*Corresponding author*: qau_abbas@yahoo.com

Inamuddin and Tariq Altalhi (eds.) Advanced Redox Flow Technology, (127–160) © 2024 Scrivener Publishing LLC

6.1 Introduction and Overview

The consumption of energy due to the rapid population growth and advancement of technology climbed to 53 billion kWh in 2020 [1]. The environment of the world is changing rapidly due to the burning of fossil fuels. To meet the escalating energy needs, renewable energy sources are widely used nowadays. The power generated from fossil fuels and renewable resources is roughly 78% and 13% ,respectively [2]. Therefore, the consumption of fossil fuels has been greatly decreased and it is shifting towards an environment-friendly energy mechanism. The population of the world is growing at an average rate of 0.9% per year [3]. Therefore, it is essential to promote an extensive power storage mechanism. The International Energy Agency has reported that there will be a 60% growth in the demand for renewable energy reserves by 2040 [4]. Due to the fluctuating nature of renewable energy resources, they did not have consistent and instant supplies, which led to the destabilization of the energy storage system. Thus, there is an essential need to develop an energy storage system that works efficiently and possesses the following characteristics: cost-effectiveness, reliability, safe, long life, maximum storage capacity, and fast response time [5]. The electrochemical storage systems, especially the secondary vanadium flow batteries, are inevitable choices due to their suitability and better performance [6]. The most common lithium-ion battery is commercially used nowadays because of its large energy density, extended life cycle, improved cycling life, and stability. Lithium-ion battery power technology is a more expensive technology than vanadium redox flow technology (VRFB). This is because vanadium is a naturally more abundant element [7]. Lead-acid battery technology has also been employed on a commercial scale. The few important properties of lead-acid battery technology are much inferior to lithium-ion batteries in life cycle, energy density, and stability. In addition, the main issue with lead-acid batteries is environmental damage. Therefore, the VRFB becomes the preferred choice in secondary storage systems with a large abundance, cost-effectiveness, high energy efficiency, long cycling life, and low toxicity [8]. It is believed that VFRB can be incorporated into other storage systems such as wind power and photovoltaic systems to fulfill the future needs of electric supply [9]. There are numerous kinds of redox flow batteries (VFRB) depending upon the redox (Oxidation-Reduction) couple utilized in batteries. Hydrogen-bromine [10], all-vanadium [11], zinc-bromine [12], all-iron [13], and semi-solid lithium [14] are the main RFB reported in the literature. Famous researchers Skyllas-Kazacos *et al.* [15] 1985 published a work in which they suggested a vanadium battery system of first generation (G-1). The work on G1 VRFB continued until the 2000s [16]. The electrochemical couple used in G1 consists of two different ionic reactions of vanadium in a diluted environment. The oxidation states of vanadium(V) involved are V^{2+}, V^{3+}, V^{4+}, V^{5} [17]. The chemical reaction of G1 is represented by the following equations.

At anode:

$$VO^{2+} + H_2O \Leftrightarrow VO^{2+} + 2H + e^{-}E^{o} = 1 \tag{6.1}$$

At cathode:

$$V^{3+} + e^{-} \Leftrightarrow V^{2+}\ E^{o} = -0.26\ V \tag{6.2}$$

The redox couple in this scheme is detached by a cationic exchange membrane which transfers the proton by avoiding the mixing of fluids. The problem of cross-contamination is very much reduced in G1 VRFB but its application is restricted due to the small specific energy of around 14.9–24.9 Wh/kg with a small temperature range (10 to 40 °C) [15]. In more recent years, investigation on VRFB has continued to explore new ways to increase the solubility of vanadium as a second generation (V/Br). In the second generation (G2), the electrolyte is a combination of vanadium bromide and vanadium chloride [18]. The positive potential of the bromide couple is lesser than that of the vanadium couple. Thus, during the charge transfer phenomena, the bromide ions will oxidize the anode. The chemical reactions in G2 are,

At positive electrode:

$$2Br^{-} + Cl^{-} \Leftrightarrow ClBr_2^{-} + 2e^{-} \tag{6.3}$$

At negative electrode:

$$2V^{3+} + 6e^{-} \Leftrightarrow 3V^{2+} \tag{6.4}$$

The main drawback of G-2 is the evolution of bromine vapors during the charge transfer process. The G2 still faces the problem of high cost rendering the use of bromine complexing agent which stops the release of bromine gas.

The third generation (G3) of VRFB was developed at State Laboratory PNNL in the United States with mixed sulfuric acid/hydrochloric acid electrolytic solutions in both the partial cells. The following improved properties were observed in G3 compared with G1 and G2: wider temperature range, faster kinetic reaction, and large stack efficiency [19].

6.1.1 Working Principle of VRFB

One of the most common and favorable battery technologies is the VRFB. The VRFB differs from other chemical storage devices in the storage of electrolyte. The flow battery contains a stack and two tanks of electrolytes. The redox flow battery stores the electrolyte in an external tank that is far away from the battery center [20]. The vanadium (V) has different oxidation states, i.e., V^{3+}, VO^{2+} (V^{4+}), VO_2^+ (V^{5+}), and V^{2+}. Each tank has a different redox couple which is indicated by the labeled Figure 6.1.

The electrolyte is supplied throughout the separate half-cells inside the battery and then recirculates over the set of cells (also called stacks) by a mechanical pump. Each half-cell contains dual plates and an electrode that are detached by a membrane whose function is to stop the mixing of electrolytes. The two half-cells constitute an electrochemical cell which is a basic unit of RFB. The cell stack assembly is designed for individual

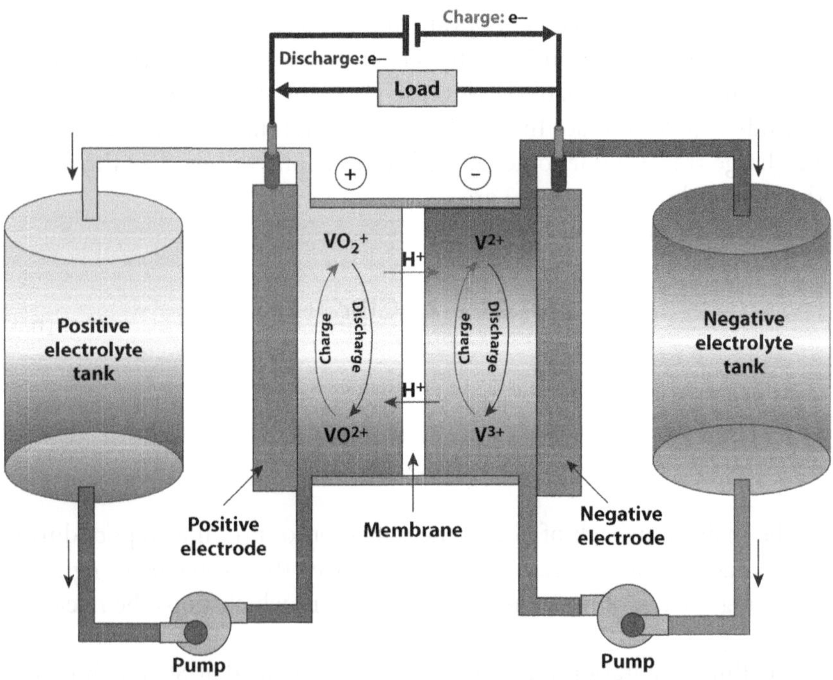

Figure 6.1 Representation of VRFB technology mechanism. Reproduced with permission [21] Copyright (2021) Elsevier.

cells. The electrolytic reaction contributes to the extension of the life of the battery [22]. The partition between the electrode and the electrolyte tank provides better power and energy for the RFBs. The complete energy stored by the battery system can be determined by its volume. The power of the battery depends upon the active materials, dynamic activities, and total area of the electrode's surface [23]. In its charging operation on the anode side, V^{4+} is oxidized to V^{5+} with the emission of an electron and proton, while on the cathode side, the trivalent ion is reduced to a divalent ion by the capture of the electron that reaches through an external circuit. The different color in the cell determines the different vanadium ions such as V^{2+}, V^{3+}, VO^{2+}, and VO_2^+ representing violet, green, blue, and light yellow, respectively [24]. The positive and negative cell reaction is represented in the equations below.

At anode cell reaction:

$$\mathrm{V\,O_2^+ + 2H^+ + e^- \Leftrightarrow VO^{2+} + H_2O + e^-} \tag{6.5}$$

At cathode cell reaction:

$$V^{2+} \Leftrightarrow V^{3+} + e^- \tag{6.6}$$

Complete reaction:

$$V\,O_2^+ + V^{2+} + 2H^+ \Leftrightarrow V\,O^{2+} + V^{3+}\,H2\,O \tag{6.7}$$

Equations (6.5) and (6.6) give the half-cell reactions which occur at the +ve and –ve electrodes. The VRFB multiple stacks are connected in a parallel manner which maintains the steady concentration of redox ions and lessened the pressure drop [25]. In its discharging operation, the negative electrode becomes V^{2+} while the positive electrode after oxidation becomes V^{5+} (VO_2^+) [26]. The positive and negative electrodes contain V^{4+} and V^{3+}, respectively. In recharging the battery, the electron transfer through bipolar plates from the positive side to the negative side diffuses positive ions from the membrane to the negative side [27]. Similarly, the reaction gets reversed during the discharging phenomenon. The development of VRFB increases its efficiency from 70% to 80% [28]. The main difference between RFB and solid-state batteries is that the RFB energy is not stored in redox material inside the electrode; instead, the energy is stored in terms of the electrolyte pumping system in tanks. These working conditions between electrolytes also enhance the battery life. The operational issue of vanadium battery such as performance, the capacity of energy storage, and efficiency directly depend on the design of the battery [29]. The design of the battery is such that their main components are separated, for example, storage tanks and stack [30].

6.1.2 Main Components of the VRFB System

6.1.2.1 Electrodes

The important parts of an RFB technology in which oxidation and reduction reactions are performed are called electrodes. The electrode surface behaves as a catalyst for chemical reactions. The surface porosity delivers the reaction place for the electrolytic solution [31]. There are two types of electrodes in VRFB; one is a positive electrode and the other is a negative electrode. A reversible reaction should be performed at the electrodes and oxidation-reduction precursors should be solvable in the electrolytic material which provides a reversible chemical reaction. The oxidation and reduction states are stable throughout the reaction. The main objective is to have a large value of redox potential difference which results in maximum electricity generation. For the better performance of flow batteries the surface area should be high, which provides a triple-phase boundary and several reaction locations [32]. To properly react, the electrode material is essential to be catalytic between the redox particles [32]. The acidic nature of electrolytes creates a corrosive atmosphere for cells. The corrosion resistance of the electrode must be large. The electrolytes are acidic such as H_2SO_4 in which the carbon and graphite electrodes show strong corrosion resistance and good electrical conductivity [32, 33]. In the design of VRFB, it is essential to take into account the geometry of electrode and electrolyte movement. Mass transport is significantly controlled by these factors and also improves battery performance. The high value of current is achieved by maximizing the current density and surface area [34]. Two of the most well-known materials utilized are graphitic poly-acrylamide and cellulose fibers in carbon felt and paper form [35]. Due to different structures and good chemical properties, carbon-based electrodes have been mainly focused on in previous years [36]. The electrode types are further subdivided into 2D and 3D electrodes. The 2D electrodes are planar and their plates are smooth while the 3D electrodes have a porous structure. The 3D carbon electrodes are carbon-based paper, felt, and cloth. These three types have different properties. The electrical resistivity of carbon paper is much lower than other 3D electrodes [36]. The high-power densities (57 MW) of carbon paper electrodes surpass any graphite felt. The disadvantage of carbon paper electrodes is their limited reaction site. To obtain large efficiency of charging-discharging, a huge reaction area is compulsory. The flow-through cell in the carbon electrode must be lower to manage charge and discharge efficiency. When the quantity of cells increases, it enhances the flow resistance, which shows a reduction in the total efficiency due to pumping power. So, the carbon electrodes are not suitable due to flow rate restrictions. The fast charge and discharge times will be required for many applications which directly rely on the rate of electrolytic flow. Hence, the carbon paper-based electrode material does not facilitate fast charge/discharge. The comparison of carbon paper with graphite felt shows that the graphite electrode has higher resistivity [36]. Graphite felt is mainly used as a negative electrode component but when we use it as a positive component, discharging occurs during the charging of battery cycles [37]. All these factors have a negative influence on the efficiency of VRFB. The graphite felt electrode seems to be the most used electrode in VRFB due to the following advantages: bolstering local, improved cell mixing mass transport, and high chemical conversion per electrolyte pass [38]. With the chemical inertness and less potential of reaction with electrolyte (H_2SO_4), the graphite felt has vast applications in VRFB [36]. The natural felt electrode has disadvantages such as poor wettability and potential degradation under long operations [39]. The presence of impurities in the electrode causes a loss in a system due to impurities reacting with surrounding material [36]. The kinetic polarization loss will be much lower in graphite felt as compared to graphite plate with single serpentine flow fields. So, the electrodes are a vital component in VRFB. The complete working of VRFB is influenced by the electrode. Different electrodes are discussed in Table 6.1.

6.1.2.1.1 Negative Electrode Material of VRFB

In the vanadium battery, the negative electrode component is made up of carbon-based materials. At the negative electrode, the redox reaction V(II)/V(III) and V(III)/V(II) occur. The performance of the battery and chemical stability of electrodes is enhanced by the alteration of carbonic electrodes. Metal oxides and heteroatoms (oxygen and nitrogen) are used to modify the electrodes [52]. Vazquez–Galven *et al.* developed porous graphite felt negative electrodes which are based on TiO_2 and nitrogen-oxygen heteroatoms [52]. The designed electrode behaves as a catalytic material for redox reactions. The prepared electrode reduces ohmic loss and increases the reversibility of redox reactions at the negative electrode. The efficiency of the modified graphite electrode was 71%. The reaction mechanism of oxidation and reduction for modified electrodes was also discussed. He *et al.* (2018) also prepared a negative electrode component (TiO_2-modified carbon fiber) by using the electrospinning technique for

Table 6.1 Different electrodes for VRFB. Reproduced with permission [40] Copyright (2020) Elsevier.

Electrodes material	Catalyst	Manufacturing materials	Summary	Reference
Graphite felt fibres Nitrogen-doped	Carbon nanospheres (NCS)	Chemical action using dopamine as nitrogen and carbon instantaneously	Good wettability with better electrochemical activity. The efficiency of the electrode is increased. It provides adaptability in a different temperature range.	[41]
Graphite-felt	Phosphorous doped	One-pot hydrothermal technique	Enhance electrochemical reaction with improved reversibility. Temperature lies between −20 to 60 °C and better stability in thousand cycles.	[42]
Graphite-felt	$KHCO_3$ carbon	Hydrothermal technique with $C_6H_{12}O_6$	Better electrolyte, improved efficiency, capacity with huge current density.	[42]
Graphite-felt	Rayon and PAN-based	Use an electric heater for thermal treatment under air	PAN-based exhibited frequent deactivation and rayon-based felt showed a continued performance.	[43]
Carbon paper	Carbon-based materials (Graphene and reduced graphene)	Electrode surface deposition	Reduce graphite oxide indicates enhanced mechanics as compared with pristine graphene.	[44]
Carbon felt, paper, rod graphite, nanofibers	HNO_3	Chemical treatment	Carbon nanofibers and carbon felt indicate less graphitization as compared to rod graphite. The carbon paper electrode showed low electrocatalytic performance with low reversibility.	[45]

(Continued)

Table 6.1 Different electrodes for VRFB. Reproduced with permission [40] Copyright (2020) Elsevier. (*Continued*)

Electrodes material	Catalyst	Manufacturing materials	Summary	Reference
Carbon nano-fibers graphite-based	graphite nanopowders	A kind of physical process	The electrode showed enhanced electrochemical activity in comparison with pristine electrodes. The vanadium constituents enhance reversibility by increasing the contents.	[46]
Carbon paper	-	KOH activation	The two-scale porous electrodes are created with improved active sites and flow rate of electrolytes. Overall, the performance of the battery is improved	[47]
Carbon felt	SnO_2 nanoparticles	Hydrothermal method	Sn shows more stability with 76% energy efficiency and 23.7% of discharge capacity.	[48]
Graphite- felt	-	Acidic reaction performed at room temperature	It indicates less peak potential separation, increased peak current densities, large charge/discharge capacity and greater efficiency.	[49]
Carbon felt	Multiwalled carbon nanotubes	Solution accumulated on multiwall carbon nanotubes	Improved coulombic and energy efficiencies	[50]
Carbon nanofiber	-	N-doping procedure	Large discharging capacity	[51]

VRFB [53]. He observed that the TiO_2 decreases the potential difference between the anion and cation peaks corresponding to V^{2+}/V^{3+} and V^{3+}/V^{2+} respectively. He also noticed the electrolytic effect of the active sides of TiO_2 on carbon fiber-based negative electrodes. The modified carbon fiber electrode shows an increase in anion and cation peak current with increasing efficiency [53]. However, the active surface area is decreased by increasing the amount of TiO_2 in the modification process which also increases resistance. Hou *et al.* (2018) also reported TiO_2 modified carbon electrodes by using hydrothermal techniques [53]. Jiang *et al.* (2019) published a paper that relates to preparing a MnO_2 nanosheet-based anode for VRFB. The anode was prepared by using the redox deposition technique under optimum conditions, i.e., time of deposition, $P^{H,}$ and amount of reactants [51]. The MnO_2 and Mn-O-H structures were formed on the carbon surface electrode which is used as the negative electrode in VRFB. The prepared electrode results in better kinetics and an increase in the discharge capacity of electrode materials [51]. The N-doped electrodes such as pyridinic N-based, quaternary N-based, and nitrogen-doped carbon show a large value of discharge capacity and large efficiency as compared to carbon electrodes [54].

6.1.2.1.2 Positive Electrode Material for VRFB

In vanadium flow technology, the positive electrode performs a dynamic part in the performance of the battery. The carbon-based electrodes are mostly used as the negative electrode in the battery. The carbon-based materials cannot be utilized as positive electrodes due to their graphitic oxidation in the process of charging [55]. For use of carbon materials as a positive electrode, it needs proper modification. The modified positive electrode increases the life cycle and performance of the battery [56]. Gursu and colleagues prepared a cathode for a vanadium flow battery system [57]. They developed the heteroatom-doped graphene-coated positive electrode by using the cyclic voltammetry technique at room temperature. They called this method Yucel's method [56]. The different functional groups such as -C-S-C-, -C-SO_x (x: 2, 3), -C-NO_2, pyrrolic N, pyridinic N, -Cl, -ClO_4, and -ClO_3 are used in doped graphene electrodes by the varying potential in voltammetric treatment [57]. These changes enhance the capacity and cyclic life of VRFB by creating electro-catalytically active kinetics on the graphene structure [58]. Le *et al.* (2016) developed a three-dimensional (3D) altered graphite-felt nano-walled structure-based electrode utilizing the chemical vapor deposition (CVD) technique [59]. The prepared electrode enhances the redox reaction rate and energy efficiency as compared to graphite felt [59]. Hosseini *et al.* (2018) prepared the -N and -WO_3 altered graphite-based positive electrodes by using a hydrothermal process [60]. The modified electrode was further characterized by spectroscopic, microscopic, and electrochemical methods. Faraji *et al.* (2017) prepared the graphite electrode by using carbon nanotubes and WO_3 composite [61]. The modified electrode showed an increase in peak current density which reduces the charge transfer resistance. Chang *et al.* (2017) discussed the making of the positive electrode of graphite felt at high temperatures (1000 °C) where CO_2 is used as an activation agent [62]. The more active sites were formed on the electrode surface by increasing the concentration of oxygen. The voltage, energy, and coulombic efficiency of the battery also increased due to these active sites [62]. Xiang with his coworkers published a work that relates to the amendment of binary $NiCoO_2$ on graphite electrodes by the combination of hydrothermal and calcination processes. The high value of discharge capacity and voltage efficiency was observed [63].

6.1.2.2 Electrolytes

6.1.2.2.1 Effect of Electrolytes on the Performance of VRFB

The electrolyte is the basic component of VRFB. The two electrolytes in the redox flow battery provide electric potential by reacting with each other. The electrolyte consists of two major parts, i.e., active redox couple (solvent) and supportive electrolyte (solvent) [64, 65]. In the vanadium flow battery, the two electrolytes use similar active redox species and it greatly decreases capacity loss and produces a cell potential of almost 1.25 V [64]. Various vanadium compounds such as vanadium tri-chloride (VCl_3) [66], vanadyl sulfate ($VOSO_4$) [67], and vanadium pentaoxide (V_2O_5) [68] are utilized in dissimilar supporting solutions which include sulfuric acid, hydrochloric acid, and sodium hydroxide to synthesize vanadium electrolyte. The solute and supporting solution have drawbacks due to which their usage is limited. The drawback to the use of HCl is to produce chlorine gas and V_2O_5 produces VCl_3 and low solubility [69]. $VOSO_4$ is initially used to prepare vanadium electrolytes [70].

The electrolyte affects the performance of VFRB as ion concentration and electrochemical activity of electrolytes influence the specific energy density of VRFB. The vanadium ion concentration and the size of the electrolyte can be modified. The extremely active electrochemical species lead to a lesser stake size [71]. The solubility of vanadium ions also affects the energy density of VFRB [32]. The precipitates of V^{+5} ions formed at temperatures>40 °C are also an issue of prime importance. This issue can be solved by using bromine, cerium, and magnesium couple. However, they all face cross-contamination issues. The temperature strength of vanadium electrolytes has been enhanced by proper modification with different acids as well as inorganic and organic supportive additives [72, 73]. The cell potential can be improved from 1.26 to 1.85 Volts in Zn-Br batteries but the Coulombic efficiency can be reduced. In VFRB hydrogen and oxygen generation cannot affect the Coulombic efficiency as compared with other battery systems. This is the main reason to consider the VRFB as a potential storage system for the future. In VRFB, cross-contamination cannot occur due to the similar ions in the half cells but this could cause the capacity fading due to the small volume of mixing. These damages can be overcome by an appropriate rebalancing tactic [74].

6.1.2.2.2 Synthesis of VRFB Electrolyte

The vanadium electrolyte is prepared from $VOSO_4$ and V_2O_5 as shown in Figure 6.2. Initially, $VOSO_4$ can be used as starting material because of its 10 times greater solubility in sulfuric acid as compared to V_2O_5 [37, 75]. The growth of VO_2^+ electrolytes with numerous sulfuric acid and vanadium concentrations was discussed in the literature [76, 77]. There is another method that starts with V_2O_5 employing chemical reduction and electrolytic dissolution [78]. In the electrolyte dissolution technique, the powder of V_2O_5 is adjourned in sulfuric acid solution at the negative half-cell and a similar proportion of sulfuric acid at the positive half-cell. During the oxidation process of water, oxygen is evolved at the positive electrode of the cell. The V_2O_5 is reduced to $V^{3.5+}$ (similar amount as VO^{2+} and V^{3+}) at the negative pole as given in the equation below.

$$V_2O_5 + 8H^+ + 3e^- \rightarrow V^{3+} + VO^{2+} + 4H_2O \tag{6.8}$$

In the process of chemical reduction, V_2O_5 in sulfuric acid has a reduction reaction to form VO^{2+} by the addition of a reducing agent (oxalic acid) as shown in the equation below.

$$V_2O_5 + (COOH)_2 + 4H^+ \rightarrow 2VO^{2+} + 2CO_2 + 3H_2O \tag{6.9}$$

For the use of VO^{2+} or $V^{3.5+}$ as an electrolyte, the following steps are required. In the case of the $V^{3.5+}$ electrolyte, it is circulated in the positive half-cell and negative half-cell with a proportion of 2:1 while the cell has been pre-charged. The process of pre-charging occurs by the VO^{2+} on the positive side is oxidized into VO_2^+ and reduced into V^{2+} on the negative

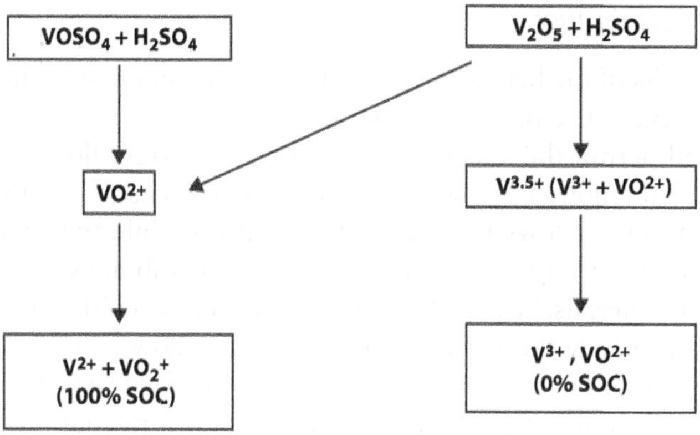

Figure 6.2 Synthesis of VRFB electrolyte. Reproduced with permission [69]. Copyright (2017) Elsevier.

side. Upon a 100% state of charge (SOC), half of the positively charged electrolyte is eliminated which is essential to balance the cell [79].

Similarly, starting with the $V^{3.5+}$ electrolyte, the same amount of electrolyte is circulated in the negative and positive tanks. The cell is initially precharged until $V^{3.5+}$ oxidizes to V(IV) on the positive side and reduces to V(III) on the negative side which corresponds to a 0% state of charge [80]. Currently, the V_2O_5 (99.8%) with high purity is used as a starting material. The research to decrease the cost of electrolytes requires more attention.

6.1.2.2.3 Electrochemical Performance of VRFB Electrolyte

The energy efficiency of a VRFB is equal to the proportion of electrical energy of the cell in discharging state and energy provided to the cell in charging. Energy efficiency can be represented by the following equation:

$$\varepsilon_{Energy} = \varepsilon_{voltage} * \varepsilon_{Coulombic}$$

Where $\varepsilon_{voltage}$ = discharging voltage/charging voltage while $\varepsilon_{Coulombic}$ = discharge capacity/charge capacity. The energy efficiency of the VRF battery system can be affected by electrolytes due to the following factors; (i) the ionic conductivity of electrolyte affects the ohmic polarization of VRFB which corresponds to the voltage efficiency, (ii) shunt current in bipolar stack lowers the energy efficiency, (iii) side reactions lower the Coulombic efficiency, and (iv) vanadium and water crossover results in the lower Coulombic efficiency due to repeating cycle.

6.1.2.3 *Membranes*

Both half-cells of the battery are separated by a component called a membrane. The main use of the membrane is to stop the interaction of electrolytes and permit the ions to move during the current flow [17]. Cation exchange and anion exchange are the two main types of membrane. The anion membrane allows the transport of anions while the cation membrane allows the transport of cations [81]. The membranes are composed of different materials, i.e., perfluorinated ionomers, nonfluorinated hydrocarbons, and partly fluorinated polymers. Many flow batteries use polymeric membranes both for anion exchange membrane (AEM) and cation exchange membrane (CEM). Among these types of membranes, the perfluorinated membrane is commercially used because of its better chemical stability and great ionic conductivity [82].

The following parameters are evaluated to be used as a membrane in a vanadium flow battery, for example area resistance, vanadium ion permeability, ion exchange capacity, ionic conductivity, thermal stability, cell performance, water transport, and chemical stability [83]. These parameters are important for membrane performance, e.g., vanadium ion permeability will permit the exchange of vanadium ions between half-cells during the operation which results in the accumulation of vanadium ions in the one-sided half-cell and depletion in the second, which reduces the overtime of the battery [84]. The periodic mixing of solution in two cells eliminates the accumulation of ions. The membrane characteristics and system operations show a significant part in the required frequency of this mixing. In the first generation, the membrane resists V^{5+} and bromine in the second generation [55]. The shifting of water from one to the other half cell is an important issue in the membrane which produces vanadium precipitates. The direction and path of water transmission rely on the nature of the membrane. In the AEM, the water is transferred from positive to negative half-cell while in the CEM reverse occurred [85]. This problem can be temporarily solved by successive use of anion exchange membrane followed by cation exchange membrane in the same multiple stacks [86]. Recently, the exchange of water as a function of the state of charge (SOC) to CEM was developed which shows that the SOC depends on the direction of water transfer. The state of current increases from 50% to 100% when water shifts from negative to positive half-cell and remains less than 50% in case of water transfer from positive to negative half-cell [80]. Hence, the key component of flow batteries is the membrane which defines the performance of the battery. The main problem in the commercialization of redox batteries is their high cost in which the weightage of the membrane is about 20% [82]. Therefore, the focus of research on the performance of membranes along with cost reduction will enhance the commercial development of RFBS.

6.1.2.3.1 Type of Membranes for VRFB

Different categories of membranes are established to meet the required properties for VRFB. The types of membranes commonly used for RFB include porous membranes and ion exchange membranes. Based on the functional groups, the ion exchange membrane is further divided into cation and anion exchange membranes. Moreover, the membrane that conducts both anion and cation is known as the amphoteric ion-exchange membrane. The ion exchange membrane consists of different layers of anion and cation. The characteristics of VRFB against different membranes are given in Table 6.2.

6.1.2.3.2 Cation Exchange Membrane (CEM)

The type of membrane which holds negative charge functional groups, i.e., $-C_6H_4O^-$, $-PO_3^{2-}$, $-PO_3H^-$, .,$-SO_3^-$ and $-COO^-$ in their structure is referred to as a CEM [92]. The carboxylic acid functional group is not feasible to make a membrane because it promotes weak exchange capacity in polymers. The sulphonic and phosphonic acid functional groups behave as durable cation exchangers because they detach movable cation effortlessly from the group and are consequently mainly applied in the making of CEM in vanadium flow batteries. However, this membrane in half-cell separation affects the energy capacity of the battery due to the cross-contamination of vanadium ions. A few studies were presented on the distribution process of vanadium ions in CEM. The most popular CEM in redox flow batteries is Dupont Nafion, developed as a copolymer of sulfonated and tetrafluoroethylene which strongly support the acid electrolytes and oxidative ions such as VO^{2+} [93]. The disadvantages of this membrane include high cost, low ion selectivity, and high permeability to vanadium ions. The cation exchange membrane has two areas with dissimilar hydrophilicity which is liable for the working of the membrane [94]. In the Nafion membrane, there is a large difference in the behaviors of the two regions. An electrically neutral, hydrophobic and flexible skeleton and SO_3- functional group, highly hydrophilic and associated with H^+ ions forming a side chain [95]. Karimi *et al.* reported that the Nafion membrane properties are highly dependent on their morphological structure. The gap between the functional group and the structure of the membrane is due to the different natures of the region, which is increased by the solvation process [96]. The

Table 6.2 Different types of membranes used in VRFB. Reproduced with permission [21]. Copyright (2021) Elsevier.

Membrane	Type	Thickness (µm)	Water uptake (Wt %)	Diffusion coefficient ($cm^2.min^{-1}$)	Current density ($mA\ cm^{-2}$)	Coulombic eff.(%)	Energy eff. (%)	Voltage eff.(%)	Ionic conductivity ($mS.cm^{-1}$)	Ref
PVDF/HA–45	CEM	-	46	$2.5*10^{-7}$	100	95	80	84	69	[87]
0.7 µm PB	AEM	30	-	-	120	98	83	85	2	[88]
SPEKS/sGO (0.5%)	CEM	50	31	$5*10^{-8}$	40	99	83	83	51	[89]
PSU/PSf-c- PTA-1.4	AEM	125	37	$2.6*10^{-7}$	120	98	84	86	20	[90]
SPPO-2%GO	CEM	40	16	$1.1*10^{-8}$	40	98	69	71	16	[91]

diameter offered by the bunch in states of hydration and balance is due to the reduction of entire free energy [97]. The sulfonated ion cluster of diameter (4 nm) is linked with a water channel of diameter (1 nm), which are interlinked and distributed inside the skeleton [98], where the transport of proton occurs in the membrane [99]. Moreover, the arrangement of this model decreases the hydrophobic interaction of water and the repulsion of the sulfonated group in the composition of the membrane [97]. The hydrophobic structure provides chemical stability while the hydrophilic structure domain acts largely for membrane conductivity [95]. The high cost of Nafion membranes limits their use while other types of membranes are also tested in VRFB. Most of the other membranes used have the sulfonated functional group, for example, phenyl sulfone [100], fluorenyl-ether ketone sulfone) [101], phthalazinone ether ketone), mixed tungstophosphoric acid [102], ether ketone composite, polypropylene and perfluoro sulfonic acid [103], ether sulfone, sulfonated ether-ether ketone) composite [104], poly(arylene ether) [105], poly(phenylene) [106], poly(ether ether ketone) and poly(1, 4-phenylene ether ether sulfone) [107].

6.1.2.3.3 Anion Exchange Membranes (AEM)

AEM is employed in VRFB where cations are used as active species [108]. The positive charge functional group present in AEM include $-SR_2^+$, $-NRH_2$, $-NH_3^+$, NR_3^+, $-PR_3^+$, and $-NR_2H$, [92] which repel the cation (Doman Exclusion) and stop the permeation of vanadium ion between cells [23]. In the CEM, the cation exchange group is strongly attached to the structure while in the case of AEM the functional groups are not strong. Chen *et al.* reported that both proton and sulfate anions are effectively shifted [109]. Many AEM has already been tested in VRFB. The characteristics like reduction in water transport, voltage, and total efficiency are improved [110]. Low vanadium ionic permeability, small area resistance, and less open circuit potential which is more stable over a long time [111] were also observed. The large value of energy efficiency and greater vanadium selectivity [112] can be achieved by testing different AEM. The low permeability of vanadium ion could be related to large Coulombic efficiency, large current efficiency, and high working stability over large cycles [109, 113, 114]. Tang *et al.* discussed that the use of selenium AEM decreases the diffusion coefficient of vanadium ions when compared with CEM [115]. They also reported that the decrease in capacity is observed in AEM over some time.

During the preparation of the membrane, the different parameters, i.e., temperature, time, and proportion of animation method also affect the characteristics of VRFB. Xing *et al.* synthesized the exchange membrane of quaternized phthalazinone ether sulfone and observed these parameters [112]. It was also noticed that the capacity of ion exchange has been enhanced with the enhancement of these three parameters with a reduction in resistance. Moreover, the AEM not only shows the increase in selective vanadium ions but the increase in efficiency was also observed as compared to Nafion membrane tests.

Sun *et al.* discussed three different anion exchange membranes with diverse ion exchange capacities for both poly (phenylene). Different anion exchange membranes are functionalized by specifically quaternary ammonium groups [116]. The effects were compared with the Nafion membrane in footings of specific resistance, and durability. The experiment shows that the lowest ion exchange capacity results in the highest specific resistance. The highest Coulombic efficiency (99%) was also observed due to the lesser crossing of VO^{2+} ions.

6.1.2.3.4 Porous-Type Membranes

The porous membranes in redox flow batteries restrict the use with very small Coulombic efficiency of VFRB and a porous membrane is now considered a separator technology. Chieng *et al.* demonstrated the porous separator in VRFB [117]. They used the modified Daramic microporous separator because of its lower price and more stable vanadium electrolyte. The Coulombic efficiency of the porous membrane is 77% and it is increased by the treatment with polyelectrolyte/ion exchange up to 90%. The porous membranes show stability even after 1650 cycles of a vanadium flow battery. Later on, Mohammad and Skyllas-Kazacos observed that the Daramic membrane is very less stable [112, 118]. The ultra and nano-filtration separator is also available with a microfiltration separator [119]. Among all these membranes, the nano-filtration membrane shows good results in VRFB. The nano-filtration membranes consist of numerous minor pores where the pressure (5 to 20 bars) was used for the saturation of ions through the membrane [120]. The separation mechanism of a nano-filtration membrane differs from an ion exchange membrane which was founded on the characteristics of changing selectivity between ion of vanadium and proton but not including the pore size [121]. The vanadium ions are separated from protons as a result of a change in radius and charge density. The charge density of the vanadium (V) ion is greater than the H^+ ion [122]. Therefore, the pores of the membrane are designed so that only H^+ ions are allowed to pass and stop the vanadium ion's cross-contamination. The main problem in the ion-exchange membrane was the permeability of vanadium ion which is much reduced by a nano-filtration membrane. Zhang and his coworkers prepared a V/H nano-filtration membrane that consists of polyacrylonitrile and the Coulombic efficiency successfully rises to 95% with an energy efficiency of almost 78% [121]. Zhang *et al.* tailored this membrane by adding silica and the results show a high increase in V/H selectivity with energy efficiency (80%) and Coulombic efficiency (97%) and energy efficiency up to 80% has been obtained [111]. Xi *et al.* used silica to alter the poly (ether sulfone) nano-filtration membranes which enhances the Coulombic efficiency above 10% and energy efficiency increased up to 7% as compared with the original porous membrane. Wei *et al.* reinforced the silica particles into a polytetrafluoroethylene medium and observed an increase in energy and Coulombic efficiency up to 80% and 93%, respectively. The performance of VRFB was improved due to silica particles, linked to the formation of nanoporous channels when silica particles fused into these membranes [123].

6.1.2.4 *Bipolar Plates*

The bipolar plates are electrically connected to adjacent cells and separated hydraulically [124]. In practice, the function of bipolar plates is to circulate the electrolyte through half cells. The conductivity range (20-5000 s m^{-1}) of the bipolar plates determines their cost. Noack *et al.* calculated that the system shows the same cost when inexpensive and small conductive plates are utilized over low conductive and additional expensive plates for continuous power yield [125]. The bipolar plates in VRFB cannot be metallic because of corrosive electrolytes. The dual plates are formed of graphite due to their larger conductivities, machinability, and resistance to erosion. The graphite-based platters are brittle and expensive to handle. The use of flow channels is an improved way to lessen the pressure across cells [126]. Further studies show that various factors show a significant part in the distribution of electrolytes such as the dimension of bipolar plates, the number of channels, and the permeability of electrodes [127]. The path performance is closely connected to the particular condition of the battery, hence it is hard to shape a design due to different configurations. The composite of carbon can halt the dissolution of graphite material from the fuel cell in the acidic electrolytes. Park *et al.* [128] observed that a plate with flake symmetrical natural graphite and nano-species shows large conductivity up to 114 S/cm, improved physical properties, and good stability in the electrolyte. A scientist, Han, and colleagues [129] made a dual polar plate derived from carbon by utilizing IrO_x-coated TiO_2 material. This new plate exhibited large energy and voltage efficiency with a 3% to 4% increase as compared to the graphite plate. Another bipolar plate was developed by using a carbon fabric/graphite hybrid which also showed a 6% increase in energy when compared with graphite [130].

6.2 VRFB System as Compared to Other Energy Storage Systems

There are many techniques such as electrical, mechanical, and thermal storage techniques that are being developed to meet energy needs. Table 6.3 describes these techniques with their advantages and disadvantages. Today, the most popular technique is the pumped hydro energy storage method. The pumped hydro energy storage mechanism covers 95% of energy storage in the world [131]. This technique is natural and had been used for many years. The cost per unit is quite low compared to modern energy sources. This storage system was started during the blackout period. This storage system has some disadvantages, e.g., vast amount of land required to build the infrastructure, environmental issues, and the high cost of construction of their technical setup [131].

Lead-acid batteries have remained useful in many fields of storage for over a century. The main advantage of a lead-acid battery is its operation over extensive temperature ranges. The disadvantage of the lead-acid battery includes hazardous lead-based material and lesser energy density. The lithium-ion battery is the most familiar and widely utilized technology among storage systems because the energy density is high with low discharge rates. The disadvantages of a lithium-ion battery include being sensitive to high temperatures and suffering from fast degradation. The sodium-sulfur battery (NaS) is a high-temperature battery. The NaS battery has a high value of power density, greater efficiency and long cycle life. The NaS battery needs thermal control to avoid risk.

Many other systems were developed but the sole method that could fulfill the need for an optimal energy storage system is currently not available. Out of all these emerging technologies, conventional batteries like the RFB are expected to meet the growing need for energy storage. The specific capacity of the RFB usually depends on the battery system, the stored electrolyte, the voltage of each cell, and the number of active particles. Moreover, power generation relies on electrode size and the behavior of active species [132]. The RFB storage system is very simple and needs little operator input for its operation. These benefits made the flow battery an exceptional energy-storing system for upcoming applications. So,

Table 6.3 Comparison of VRFB with different technologies. Reproduced with permission [133]. Copyright (2020) MDPI.

Electrochemical energy storage	Maximum power (Watt)	Discharge time	Maximum lifetime	Specific energy (Wh/L)	Efficiency	Capital budget (Per KW)
hydro pumping storage	$1000*10^6$	4-12 hours	30-60 years	0.2 - 2	70%-85%	165$
air compression	$1000*10^6$	2-30 hours	20-40 years	2 - 6	40%-70%	105$
Fly-wheel	$20*10^6$	Sec-minutes	20,000-100,000	20 - 80	70%-95%	11,520$
Lead acid-based battery storage	$100*10^6$	1 minute-8 hours	6-40 years	50 - 80	80%-90%	1040$
Sodium-sulfur battery storage	$100*10^6$	1 minute-8 hours	2,500-4,400	150 - 300	70%-90%	2644$
Lithium-ion battery (LIBS)	$100*10^6$	1 minute-8 hours	1,000-10,000	200 - 400	85%-95%	1084$
Redox flow Battery (RFB)	$100*10^6$	Hours	12,000-14,000	20 - 70	60%-85%	2200$

because of the amalgamation of all these characteristics, the RFB system has comparatively low capital costs and small running expenses, particularly related to other evolving energy storage techniques. Among all these flow batteries, the vanadium flow batteries are most suitable due to the following advantages: (I) no solution contamination across a membrane with diffusion of vanadium ion, (II) ions are regenerated over normal battery operation, and (III) moderate efficiency lies between 70-90% [28].

6.3 Recent Research on VRFB

6.3.1 Positive and Negative Electrodes

Numerous experiments are being performed and research is going on to improve the performance of electrodes, especially the design aspect, increase reaction sites and decrease ohmic resistance to enhance electrochemical performance. Recent research exhibits that the electrodes of carbon paper and carbon cloth show better performance. Liu *et al.* got good results when they paired up the carbon paper/cloth with zero-gap cell construction [134]. Zero gaps referred to the configuration in which the current collector, both electrodes and membrane, are in direct contact [134]. The zero-gap configuration provides large power and current densities. The small surface activation is the main drawback of this configuration which limits the current [34]. Researchers are trying to modify the surface to charge material electrical conduction. Pezeshki *et al.* applied the above techniques to increase the oxygen content and observed better performance of the cell. The heat treatment along with oxygen content increased the cell efficiency by up to 13% [135]. The proper modification in cell design also increases cell performance. The main loss of electrodes is through flow resistance and more research efforts are required to minimize the loss. The addition of flow channels can increase efficiency and reduce pumping power requirements [136]. The latest electrode materials include carbon nanotubes (CNT), carbon nano-fibers, and graphene oxide (GO). He *et al.* fixed the ZrO_2 nanoparticles into the skeleton of the nanofiber [155]. The prepared electrode demonstrated enhanced catalytic properties. Furthermore, the results showed that the voltage and current efficiencies had also improved up to 8% and remained sustainable over 50 charge/discharge cycles [137]. The resistance between the electrode and collector is affected by the cell assembly. The distribution of pressure over the cell is ignored and enhances the current density with a low-pressure drop [138]. Qu *et al.* evaluated the circulation of compression over the cell and the properties of the performance [139]. Different surface techniques and methods are used for various electrodes to increase the performance of the cell. Continued research is being carried out to improve the electrochemical efficiency of electrodes by chemical treatment. The decrease in cell resistance was witnessed with growth in acid concentration (up to 98%). The H_2SO_4 and HNO_3 were used for the chemical treatment of carbon paper electrodes through ultrasonic mixing. The alterations in the functional groups present on the carbon surface paper were around 3.8% to 14.3% [140]. Zhang *et al.* prepared the electrode material in which the gradually porous system of carbon nanofibers was bound to carbon felt. The results demonstrated a huge energy efficiency of 82.4% with a current density of 320 mA/cm^2 [141].

6.3.2 Electrolytes

The concentration of vanadium in VRFB is not more than 2M [162]. This limits the solubility of vanadium in electrolytes [66]. Kazacos *et al.* observed that the amount of sulfuric acid and vanadium ions above 40°C should be 1.5M and 3-4M respectively [142]. The vanadium ion V^{5+} form precipitates with sulfuric acid when kept for a long at a high temperature. These three vanadium ions V^{2+}, V^{3+}, and V^{4+} have much less solubility in sulfuric acid at low temperatures up to 5 °C [135]. The vanadium ion V^{5+} forms V_2O_5 above 40 °C [143] while the vanadium ions V^{2+}, V^{3+}, and V^{4+} form the sulfates, which give precipitates under 10 °C [73].

If the temperature of the electrolyte remains modest and the battery is continuously cycled then the concentration of vanadium enhances up to 3M, which raises the energy density to 35 Wh kg^{-1}. Mohamed *et al.* observed that a higher sulfuric acid concentration increases the electrical conductivity of electrolytes [144]. Rahman *et al.* revealed that the use of supersaturated positive electrolytes limits the concentration of electrolytes up to 3.5M and 5-6M sulfate/bisulfate [76]. The four oxidation states of vanadium restrict the higher concentration in the electrolyte. Moore *et al.* discussed the fact that the battery suffers a 100% discharge and more cycles will be expected between 20% and 80%, by the use of some larger concentrations of electrolyte [145].

6.4 Conclusion and Perspective

The VRFB shows good results in energy storage applications. However, some changes are still needed to progress the capacity of VRFB. In this chapter the working principle and main components of VRFB are addressed to provide a detailed study for the reader. The chapter provides an outlook from the making of batteries to recent advanced technologies. The operating system of VRFB is also discussed in detail. Technologies to prepare the electrolyte have been improved which results in better chemical stability, higher concentration, and superior current density. Different additives are proposed with electrolytes to improve the overall performance of the battery. Graphite-based electrodes were mainly used for VRFB despite their limitations. The new materials are also suggested that have greater catalytic activity, low resistance and are cost-effective. Thermal, surface, and chemical treatments are used to increase the working of the cell. The permeability of vanadium ion is the major challenge in the development of membranes which also affects the overall cost of the battery. The researchers are trying to design new membranes with smaller cross-contamination that are economical to use. Recently, various structures have been designed with different layers and nanoslits which help to drop the permeability of vanadium ions.

References

1. Wang, S., Owusu, K.A., Mai, L., Ke, Y., Zhou, Y., Hu, P., Magdassi, S., Long, Y., Vanadium dioxide for energy conservation and energy storage applications: Synthesis and performance improvement. *Appl. Energy*, 211, 200, 2018.
2. U.S. Energy Information Administration (EIA), *U.S. Energy Facts Explained*, Washington DC, USA, 2019.
3. International Energy Agency (IEA). *World Energy Outlook 2017*. https://www.iea.org/reports/world-energy-outlook-2017.
4. International Energy Agency (IEA). *World Energy Outlook 2019*. https://www.iea.org/reports/world-energy-outlook-2019.
5. Ibrahim, H., Younès, I., Perron, A.U., Energy storage systems—characteristics comparisons. *Renew. Sustain. Energy Rev.*, 12, 1221-1250, 2008.
6. Das, C.K., Bass, O., Kothapalli, G., Mahmoud, T.S., Habibi, D., Optimal placement of distributed energy storage systems in distribution networks using artificial bee colony algorithm. *Appl. Energy*, 232, 212, 2018.
7. Messaggi, M., Canzi, P., Mereu, R., Baricci, A., Inzoli, F., Casalegno, A., Zago, M., Analysis of flow field design on vanadium redox flow battery performance: Development of 3d computational fluid dynamic model and experimental validation. *Appl. Energy*, 228, 1057, 2018.
8. Wei, Z., Zhao, J., Ji, D., Tseng, K., A multi-timescale estimator for battery state of charge and capacity dual estimation based on an online identified model. *Appl. Energy*, 204, 1264, 2017.

9. Zhang, Y., Campana, P.E., Yang, Y., Stridh, B., Lundblad, A., Yan, J., Energy flexibility from the consumer: Integrating local electricity and heat supplies in a building. *Appl. Energy*, 223, 430, 2018.
10. Lin, G., Chong, P.Y., Yarlagadda, V., Nguyen, T., Wycisk, R., Pintauro, P., Bates, M., Mukerjee, S., Tucker, M., Weber, A., Advanced hydrogen-bromine flow batteries with improved efficiency, durability and cost. *J. Electrochem. Soc.*, 163, A5049, 2015.
11. Ulaganathan, M., Aravindan, V., Yan, Q., Madhavi, S., Kazacos, M., Lim, T.M., Recent advancements in all-vanadium redox flow batteries. *Adv. Mater.*, 3, 1500309, 2016.
12. Yang, J.H., Yang, H.S., Ra, H.W., Shim, J., Jeon, J.-D., Effect of a surface active agent on performance of zinc/bromine redox flow batteries: Improvement in current efficiency and system stability. *J. Power Sources*, 275, 294, 2015.
13. Hawthorne, K.L., Petek, T.J., Miller, M.A., Wainright, J.S., Savinell, R.F., An investigation into factors affecting the iron plating reaction for an all-iron flow battery. *J. Electrochem. Soc.*, 162, A108, 2014.
14. Duduta, M., Ho, B., Wood, V.C., Limthongkul, P., Brunini, V.E., Carter, W.C., Chiang, Y.M., Semi-solid lithium rechargeable flow battery. *Adv. Mater.*, 1, 511, 2011.
15. Sum, E., Kazacos, M., A study of the V(ii)/V(iii) redox couple for redox flow cell applications. *J. Power Sources*, 15, 179, 1985.
16. Tang, A., Dynamic modelling and simulation of the all-vanadium redox flow battery, PhD thesis, Sydney, Australia: The University of New South Wales, 2013.
17. Eckroad, S., Palo Alto, Ca., Vanadium redox flow batteries: An in-depth analysis. *Intl. J. Energy Res.*, 39, 1014836, 2007.
18. Vafiadis, H., Kazacos, M., Evaluation of membranes for the novel vanadium bromine redox flow cell. *J. Membr. Sci.*, 279, 394, 2006.
19. Zhang, C., Zhao, T., Xu, Q., An, L., Zhao, G., Effects of operating temperature on the performance of vanadium redox flow batteries. *Appl. Energy*, 155, 349, 2015.
20. Bartolozzi, M., Development of redox flow batteries. A historical bibliography. *J. Power Sources*, 27, 219, 1989.
21. Vinco, J.H., Domingos, A.E.E., Espinosa, D.C.R., Tenório, J.S., Baltazar, M., Unfolding the vanadium redox flow batteries: An indeep perspective on its components and current operation challenges. *J. Energy Storage*, 43, 103180, 2021.
22. Ng, A., Nathwani, J., *Paths to Sustainable Energy*, IntechOpen, 2010.
23. Ye, R., Henkensmeier, D., Yoon, S.J., Huang, Z., Kim, D.K., Chang, Z., Kim, S., Chen, R., Storage, Redox flow batteries for energy storage: A technology review. *J. Electrochem. Energy Convers. Storage*, 15, 010801, 2018.
24. Kear, G., Shah, A.A., Walsh, F.C., Development of the all-vanadium redox flow battery for energy storage: A review of technological, financial and policy aspects. *Intl. J. Energy Res.*, 36, 1105, 2012.

25. Arenas, L., De León, C.P., Walsh, F., Engineering aspects of the design, construction and performance of modular redox flow batteries for energy storage. *J. Energy Storage*, 11, 119, 2017.
26. Gundlapalli, R., Kumar, S., Jayanti, S., Stack design considerations for vanadium redox flow battery. *INAE Lett.*, 3, 149, 2018.
27. De León, C.P., Ferrer, A.F., García, J.G., Szánto, D.A., Walsh, F., Redox flow cells for energy conversion. *J. Power Sources* 160, 716-732, 2006.
28. Soloveichik, G., Flow batteries: Current status and trends. *Chem. Rev.*, 115, 11533, 2015.
29. Tang, A., Bao, J., Kazacos, M., Studies on pressure losses and flow rate optimization in vanadium redox flow battery. *J. Power Sources*, 248, 154, 2014.
30. Barelli, L., Bidini, G., Ottaviano, P.A., Pelosi, D., Vanadium redox flow batteries application to electric buses propulsion: Performance analysis of hybrid energy storage system. *J. Energy Storage, 24,* 100770, 2019.
31. Couper, A.M., Pletcher, D., Walsh, F.C., Electrode materials for electrosynthesis. *Chem. Rev.*, 90, 837, 1990.
32. Wang, W., Luo, Q., Li, B., Wei, X., Li, L., Yang, Z., Recent progress in redox flow battery research and development. *Adv. Funct. Mater.*, 23, 970, 2013.
33. Flox, C., Skoumal, M., Garcia, J., Andreu, T., Morante, J., Strategies for enhancing electrochemical activity of carbon-based electrodes for all-vanadium redox flow batteries. *Appl. Energy*, 109, 344, 2013.
34. Rivera, F.F., De León, C.P., Walsh, F.C., Nava, J.L., The reaction environment in a filter-press laboratory reactor: The FM01-LC flow cell. *Intl. J. Energy Res.*, 161, 436, 2015.
35. Eifert, L., Banerjee, R., Jusys, Z., Zeis, R., Characterization of carbon felt electrodes for vanadium redox flow batteries: Impact of treatment methods. *J. Electrochem. Soc.*, 165, A2577, 2018.
36. Castañeda, L.F., Walsh, F.C., Nava, J.L., De León, C., Graphite felt as a versatile electrode material: Properties, reaction environment, performance and applications. *Electrochim. Acta*, 258, 1115, 2017.
37. Kazacos, M., Grossmith, F., Efficient vanadium redox flow cell. *J. Electrochem Soc.*, 134, 2950, 1987.
38. Garcia, J., Bonete, P., Expósito, E., Montiel, V., Aldaz, A., Maciá, R., Characterization of a carbon felt electrode: Structural and physical properties. *J. Mater. Chem.*, 9, 419, 1999.
39. Liu, H., Xu, Q., Yan, C., Qiao, Y., Corrosion behavior of a positive graphite electrode in vanadium redox flow battery. *Electrochim. Acta*, 56, 8783, 2011.
40. Aberoumand, S., Woodfield, P., Shabani, B., Dao, D.V., Advances in electrode and electrolyte improvements in vanadium redox flow batteries with a focus on the nanofluidic electrolyte approach. *Phys. Reports*, 881, 1, 2020.
41. Wu, L., Shen, Y., Yu, L., Xi, J., Qiu, X.., Boosting vanadium flow battery performance by nitrogen-doped carbon nanospheres electrocatalyst. *Nano Energy*, 28, 19, 2016.

42. Zhao, C., Li, Y., He, Z., Jiang, Y., Li, L., Jiang, F., Zhou, H., Zhu, J., Meng, W., Wang, L., $KHC0_3$ activated carbon microsphere as excellent electrocatalyst for Vo^{2+}/Vo^{2+} redox couple for vanadium redox flow battery. *J. Energy Chem.*, 29, 103, 2019.
43. Mazúr, P., Mrlik, J., Pocedic, J., Vrána, J., Dundálek, J., Kosek, J., Bystron, T., Effect of graphite felt properties on the long-term durability of negative electrode in vanadium redox flow battery. *J. Power Sources*, 414, 354, 2019.
44. Aaron, D., Yeom, S., Kihm, K.D., Gandomi, Y.A., Ertugrul, T., Mench, M.M., Kinetic enhancement via passive deposition of carbon-based nanomaterials in vanadium redox flow batteries. *J. Power Sources*, 366, 241, 2017.
45. Di Blasi, A., Di Blasi, O., Briguglio, N., Aricò, A.S., Sebastián, D., Lázaro, M., Monforte, G., Antonucci, V., Investigation of several graphite-based electrodes for vanadium redox flow cell. *J. Power Sources*, 227, 15, 2013.
46. Wei, G., Jing, M., Fan, X., Liu, J., Yan, C., A new electrocatalyst and its application method for vanadium redox flow battery. *J. Power Sources*, 287, 81, 2015.
47. Zhou, X., Zeng, Y., Zhu, X., Wei, L., Zhao, T., A high-performance dual-scale porous electrode for vanadium redox flow batteries. *J. Power Sources*, 325, 329, 2016.
48. Mehboob, S., Ali, G., Shin, H.-J., Hwang, J., Abbas, S., Chung, K.Y., Ha, H.Y., Enhancing the performance of all-vanadium redox flow batteries by decorating carbon felt electrodes with SnO_2 nanoparticles. *Appl. Energy*, 229, 910, 2018.
49. Jiang, H., Shyy, W., Ren, Y., Zhang, R., Zhao, T., A room-temperature activated graphite felt as the cost-effective, highly active and stable electrode for vanadium redox flow batteries. *Appl. Energy*, 233, 544, 2019.
50. Wei, G., Jia, C., Liu, J., Yan, C., Carbon felt supported carbon nanotubes catalysts composite electrode for vanadium redox flow battery application. *J. Power Sources*, 220, 185, 2012.
51. Jiang, Y., Feng, X., Cheng, G., Li, Y., Li, C., He, Z., Zhu, J., Meng, W., Zhou, H., Dai, L., Electrocatalytic activity of MnO_2 nanosheet array-decorated carbon paper as superior negative electrode for vanadium redox flow batteries. *Electrochim. Acta*, 322, 134754, 2019.
52. Galván, J., Flox, C., Jervis, J., Jorge, A., Shearing, P., Morante, J., High-power nitrided TiO_2 carbon felt as the negative electrode for all-vanadium redox flow batteries. *Carbon*, 148, 91, 2019.
53. Hou, B., Cui, X., Chen, Y., *In situ* TiO_2 decorated carbon paper as negative electrode for vanadium redox battery. *Solid State Ion.*, 325, 148, 2018.
54. He, Z., Li, M., Li, Y., Wang, L., Zhu, J., Meng, W., Li, C., Zhou, H., Dai, L., Electrospun nitrogen-doped carbon nanofiber as negative electrode for vanadium redox flow battery. *Appl. Surf. Sci.*, 469, 423, 2019.
55. Parasuraman, A., Lim, T.M., Menictas, C., Kazacos, M., Review of material research and development for vanadium redox flow battery applications. *Electrochim. Acta*, 101, 27, 2013.

56. Gürsu, H., Gençten, M., Şahin, Y., Preparation of n-doped grapheme-based electrode via electrochemical method and its application in vanadium redox flow battery. *J. Energy Res.*, 42, 3851, 2018.
57. Gursu, H., Gencten, M., Sahin, Y., Preparation of sulphur-doped graphene-based electrodes by cyclic voltammetry: A potential application for vanadium redox flow battery. *Intl. J. Electrochem. Sci.*, 13, 875, 2018.
58. Gürsu, H., Gencten, M., Şahin, Y., One-step electrochemical preparation of graphene-coated pencil graphite electrodes by cyclic voltammetry and their application in vanadium redox batteries. *Electrochim. Acta*, 243, 239, 2017.
59. Li, W., Zhang, Z., Tang, Y., Bian, H., Ng, T.W., Zhang, W., Lee, C.S., Graphene-nanowall-decorated carbon felt with excellent electrochemical activity toward Vo^{2+}/Vo^{2+} couple for all vanadium redox flow battery. *Adv. Sci.*, 3, 1500276, 2016.
60. Hosseini, M.G., Mousavihashemi, S., López, S., Flox, C., Andreu, T., Morante, J.R., High-power positive electrode based on synergistic effect of N- and WO_3-decorated carbon felt for vanadium redox flow batteries. *Carbon*, 136, 444, 2018.
61. Faraji, M., Hassanzadeh, A., Mohseni, M., Interlaced WO_3-carbon nanotube nanocomposite electrodeposited on graphite as a positive electrode in vanadium redox flow battery. *Thin Solid Films*, 642, 188, 2017.
62. Chang, Y.-C., Chen, J.-Y., Kabtamu, D.M., Lin, G.-Y., Hsu, N.-Y., Chou, Y.-S., Wei, H.-J., Wang, C.-H., High efficiency of CO_2-activated graphite felt as electrode for vanadium redox flow battery application. *J. Power Sources*, 364, 1, 2017.
63. Xiang, Y., Daoud, W., Binary $NiCoO_2^-$ modified graphite felt as an advanced positive electrode for vanadium redox flow batteries. *J. Mater. Chem. A.*, 7, 5589, 2019.
64. Lourenssen, K., Williams, J., Ahmadpour, F., Clemmer, R., Tasnim, S., Vanadium redox flow batteries: A comprehensive review. *J. Energy Storage*, 25, 100844, 2019.
65. Milshtein, J.D., Drake, J., Perry, M.L., Brushett, F.R., The critical role of supporting electrolyte selection on flow battery cost. *J. Electrochem. Soc.,* 164 A3883, 2017.
66. Kazacos, M., Novel vanadium chloride/polyhalide redox flow battery. *J. Power Sources*, 124, 299, 2003.
67. Rychcik, M., Kazacos, M., Characteristics of a new all-vanadium redox flow battery. *J. Power Sources*, 22, 59-67, 1988.
68. Shibata, A., Sato, K., Development of vanadium redox flow battery for electricity storage. *Power Eng. J.*, 13, 130, 1999.
69. Choi, C., Kim, S., Kim, R., Choi, Y., Kim, S., Jung, H.-Y., Yang, J.H., Kim, H.-T., A review of vanadium electrolytes for vanadium redox flow batteries. *Renew. Sustain. Energy Rev.*, 69, 263, 2017.

70. Li, W., Zaffou, R., Sholvin, C.C., Perry, M.L., She, Y., Vanadium redox-flow-battery electrolyte preparation with reducing agents. *Electrochem. Soc.*, 53, 93, 2013.
71. Kazacos, M., Peng, C., Cheng, M.J.E., Letters, S.-S., Evaluation of precipitation inhibitors for supersaturated vanadyl electrolytes for the vanadium redox battery. *Electrochem. Solid State Lett.*, 2, 121, 1998.
72. Li, L., Kim, S., Wang, W., Vijayakumar, M., Nie, Z., Chen, B., Zhang, J., Xia, G., Hu, J., Graff, G., A stable vanadium redox-flow battery with high energy density for large-scale energy storage. *Adv. Energy Mater.*, 1, 394, 2011.
73. Roe, S., Menictas, C., Kazacos, M., A high energy density vanadium redox flow battery with 3M vanadium electrolyte. *J. Electrochem. Soc.*, 163, A5023, 2015.
74. Minke, C., Turek, T., Materials, system designs and modelling approaches in techno-economic assessment of all-vanadium redox flow batteries–A review. *J. Power Sources*, 376, 66, 2018.
75. Oei, D.-G., Permeation of vanadium cations through anionic and cationic membranes. *J. Appl. Electrochem.*, 15, 231, 1985.
76. Rahman, F., Kazacos, M., Vanadium redox battery: Positive half-cell electrolyte studies. *J. Power Sources*, 189, 1212, 2009.
77. Zhao, P., Zhang, H., Zhou, H., Chen, J., Gao, S., Yi, B., Characteristics and performance of 10 kW class all-vanadium redox-flow battery stack. *J. Power Sources*, 162, 1416, 2006.
78. Kazacos, M., Mcdermott, R.J., Vanadium Compound Dissolution Processes, 1988, Google Patents.
79. Kazacos, M., Rychick, M., Robins, R., *All-Vanadium Redox Battery*. 1988, Google Patents.
80. Sukkar, T., Kazacos, M., Water transfer behaviour across cation exchange membranes in the vanadium redox battery. *J. Membr. Sci.*, 222, 235, 2003.
81. Xu, T., Ion exchange membranes: State of their development and perspective. *J. Membr. Sci.*, 263, 1, 2005.
82. Prifti, H., Parasuraman, A., Winardi, S., Lim, T.M., Kazacos, M., Membranes for redox flow battery applications. *Membranes*, 2, 275, 2012.
83. Teng, X., Zhao, Y., Xi, J., Wu, Z., Qiu, X., Chen, L., Nafion/organic silica modified TiO_2 composite membrane for vanadium redox flow battery via *in situ* sol–gel reactions. *J. Membr. Sci.*, 341, 149, 2009.
84. Sukkar, T., Kazacos, M., Membrane stability studies for vanadium redox cell applications. *J. Appl. Electrochem.*, 34, 137, 2004.
85. Mohammadi, T., Chieng, S., Kazacos, M.S., Water transport study across commercial ion exchange membranes in the vanadium redox flow battery. *J. Membr. Sci.*, 133, 151, 1997.
86. Pellegri, A., Broman, B.M., *Redox Flow Battery System and Cell Stack*. 2002, Google Patents.

87. Rajput, A., Khan, H., Raj, S.K., Kothandaraman, R., Kulshrestha, V., Styrene-co-DVB grafted PVDF proton exchange membranes for vanadium redox flow battery applications. *Mater. Adv.*, 1, 1930, 2020.
88. Gubler, L., Vonlanthen, D., Schneider, A., Oldenburg, F., Composite membranes containing a porous separator and a polybenzimidazole thin film for vanadium redox flow batteries. *J. Electrochem. Soc.*, 167, 100502, 2020.
89. Aziz, M.A., Shanmugam, S., Sulfonated graphene oxide-decorated block copolymer as a proton-exchange membrane: Improving the ion selectivity for all-vanadium redox flow batteries. *J. Mater. Chem. A.*, 6, 17740, 2018.
90. Xing, Y., Liu, L., Wang, C., Li, N., Side-chain-type anion exchange membranes for vanadium flow battery: Properties and degradation mechanism. *J. Mater. Chem. A.*, 6, 22778, 2018.
91. Thong, P.T., Sadhasivam, T., Lim, H., Jin, C.-S., Ryi, S.-K., Park, W., Kim, H.T., Roh, S.-H., Jung, H.-Y., High oxidizing stability and ion selectivity of hybrid polymer electrolyte membrane for improving electrochemical performance in vanadium redox flow battery. *J. Electrochem. Soc.*, 165, A2321, 2018.
92. Yuan, X.Z., Song, C., Platt, A., Zhao, N., Wang, H., Li, H., Fatih, K., Jang, D., A review of all-vanadium redox flow battery durability: Degradation mechanisms and mitigation strategies. *Intl. J. Energy Res.*, 43, 6599, 2019.
93. Wang, W., Kizewski, J.P., Duan, W., *Redox flow batteries*, 2017, CRC Press, p. 43.
94. Mauritz, K.A., Moore, R.B., State of understanding of Nafion. *Chem. Rev.*, 104, 4535, 2004.
95. Ding, C., Zhang, H., Li, X., Liu, T., Xing, F., Vanadium flow battery for energy storage: Prospects and challenges. *J. Phys. Chem. Lett.*, 4, 1281, 2013.
96. Karimi, M.B., Mohammadi, F., Hooshyari, K., Recent approaches to improve Nafion performance for fuel cell applications: A review. *Intl. J. Hydr. Energy*, 44, 28919, 2019.
97. Hsu, W.Y., Gierke, T.D., Ion transport and clustering in Nafion perfluorinated membranes. *J. Membr. Sci.*, 13, 307, 1983.
98. Bordín, S.F., Andrada, H.E., Carreras, A.C., Castellano, G.E., Oliveira, R.G., Josa, V.G., Nafion membrane channel structure studied by small-angle x-ray scattering and monte carlo simulations. *Polymer*, 155, 58, 2018.
99. Yin, C., Li, J., Zhou, Y., Zhang, H., Fang, P., He, C., Phase separation and development of proton transport pathways in metal oxide nanoparticle/nafion composite membranes during water uptake. *J. Phys. Chem. C.*, 122, 9710, 2018.
100. Kim, S., Yan, J., Schwenzer, B., Zhang, J., Li, L., Liu, J., Yang, Z.G., Hickner, M., Cycling performance and efficiency of sulfonated poly (sulfone) membranes in vanadium redox flow batteries. *Electrochem. Commun.*, 12, 1650, 2010.

101. Yin, B., Li, Z., Dai, W., Wang, L., Yu, L., Xi, J., Highly branched sulfonated poly(fluorenyl ether ketone sulfone)s membrane for energy efficient vanadium redox flow battery. *J. Power Sources*, 285, 109, 2015.
102. Wang, N., Yu, J., Zhou, Z., Fang, D., Liu, S., Liu, Y., SPPEK/TPA composite membrane as a separator of vanadium redox flow battery. *J. Membr. Sci.*, 437, 114, 2013.
103. Jia, C., Liu, J., Yan, C., A multilayered membrane for vanadium redox flow battery. *J. Power Sources*, 203, 190, 2012.
104. Ling, X., Jia, C., Liu, J., Yan, C., Preparation and characterization of sulfonated poly (ether sulfone)/sulfonated poly (ether ether ketone) blend membrane for vanadium redox flow battery. *J. Membr. Sci.*, 415, 306, 2012.
105. Chen, D., Kim, S., Li, L., Yang, G., Hickner, M., Stable fluorinated sulfonated poly (arylene ether) membranes for vanadium redox flow batteries. *RSC Adv.*, 2, 8087, 2012.
106. Fujimoto, C., Kim, S., Stains, R., Wei, X., Li, L., Yang, Z.G., Vanadium redox flow battery efficiency and durability studies of sulfonated Diels Alder poly (phenylene)s. *Electrochem. Commun.*, 20, 48, 2012.
107. Macksasitorn, S., Changkhamchom, S., Sirivat, A., Siemanond, K., Sulfonated poly (ether ether ketone) and sulfonated poly (1, 4-phenylene ether ether sulfone) membranes for vanadium redox flow batteries. *High Perform. Polym.*, 24, 603, 2012.
108. Kreuer, K.-D., Ion conducting membranes for fuel cells and other electrochemical devices. *Chem. Mater.*, 26, 361, 2014.
109. Chen, D., Hickner, M.A., Agar, E., Kumbur, E.C., Selective anion exchange membranes for high coulombic efficiency vanadium redox flow batteries. *Electrochem. Commun.*, 26, 37, 2013.
110. Hwang, G.-J., Ohya, H., Crosslinking of anion exchange membrane by accelerated electron radiation as a separator for the all-vanadium redox flow battery. *J. Membr. Sci.*, 132, 55, 1997.
111. Qiu, J., Li, M., Ni, J., Zhai, M., Peng, J., Xu, L., Zhou, H., Li, J., Wei, G., Preparation of ETFE-based anion exchange membrane to reduce permeability of vanadium ions in vanadium redox battery. *J. Membr. Sci.*, 297, 174, 2007.
112. Xing, D., Zhang, S., Yin, C., Zhang, B., Jian, X., Effect of amination agent on the properties of quaternized poly (phthalazinone ether sulfone) anion exchange membrane for vanadium redox flow battery application. *J. Membr. Sci.*, 354, 68, 2010.
113. Zhang, S., Yin, C., Xing, D., Yang, D., Jian, X., Preparation of chloromethylated/quaternized poly (phthalazinone ether ketone) anion exchange membrane materials for vanadium redox flow battery applications. *J. Membr. Sci.*, 363, 243, 2010.
114. Tian, B., Yan, C.-W., Wang, F.-H., Modification and evaluation of membranes for vanadium redox battery applications. *J. Membr. Sci.*, 34, 1205, 2004.

115. Tang, A., Bao, J., Kazacos, M.., Dynamic modelling of the effects of ion diffusion and side reactions on the capacity loss for vanadium redox flow battery. *J. Power Sources*, 196, 10737, 2011.
116. Sun, C.-N., Tang, Z., Belcher, C., Zawodzinski, T.A., Fujimoto, C., Evaluation of Diels–Alder poly (phenylene) anion exchange membranes in all-vanadium redox flow batteries. *Electrochem. Commun.*, 43, 63, 2014.
117. Chieng, S., Kazacos, M.., Modification of daramic, microporous separator, for redox flow battery applications. *J. Membr. Sci.*, 75, 81, 1992.
118. Mohammadi, T., Kazacos, M., Preparation of sulfonated composite membrane for vanadium redox flow battery applications. *J. Membr. Sci.*, 107, 35, 1995.
119. Koros, W., Ma, Y., Shimidzu, T.J.P., Terminology for membranes and membrane processes, IUPAC recommendations, *Pure & Appl. Chem.*, 68, 1479, 1996.
120. Vandezande, P., Gevers, L.E., Vankelecom, I.F., Solvent resistant nanofiltration: Separating on a molecular level. *Chem. Soc. Rev.*, 37, 365, 2008.
121. Zhang, H., Zhang, H., Li, X., Mai, Z., Zhang, J., Nanofiltration (nf) membranes: The next generation separators for all vanadium redox flow batteries? *Energy Envtal. Sci.*, 4, 1676, 2011.
122. Oriji, G., Katayama, Y., Miura, T.., Investigations on V(iv)/V(v) and V(ii)/V(iii) redox reactions by various electrochemical methods. *J. Power Sources*, 139, 321, 2005.
123. Wei, X., Nie, Z., Luo, Q., Li, B., Chen, B., Simmons, K., Sprenkle, V., Wang, W., Nanoporous polytetrafluoroethylene/silica composite separator as a high-performance all-vanadium redox flow battery membrane. *Adv. Energy Mater.*, 3, 1215, 2013.
124. Lee, N.J., Lee, S.-W., Kim, K.J., Kim, J.-H., Park, M.-S., Jeong, G., Kim, Y.-J., Byun, D., Development of carbon composite bipolar plates for vanadium redox flow batteries. Bull. *Korean Chem. Soc.*, 33, 3589, 2012.
125. Noack, J., Wietschel, L., Roznyatovskaya, N., Pinkwart, K., Tübke, J., Techno-economic modeling and analysis of redox flow battery systems. *Energies*, 9, 627, 2016.
126. Tian, C.-H., Chein, R., Hsueh, K.-L., Wu, C.-H., Tsau, F.-H., Design and modeling of electrolyte pumping power reduction in redox flow cells. *Rare Metals*, 30, 16, 2011.
127. Kumar, S., Jayanti, S., Effect of flow field on the performance of an all-vanadium redox flow battery. *J. Power Sources*, 307, 782, 2016.
128. Park, M., Jung, Y.-J., Ryu, J., Cho, J., Material selection and optimization for highly stable composite bipolar plates in vanadium redox flow batteries. *J. Mater. Chem. A.*, 2, 15808, 2014.
129. Han, J., Yoo, H., Kim, M., Lee, G., Choi, J., High-performance bipolar plate of thin irox-coated $Ti0_2$ nanotubes in vanadium redox flow batteries. *Catal. Today*, 295, 132, 2017.

130. Kim, K.H., Kim, B.G., Development of carbon composite bipolar plate (bp) for vanadium redox flow battery. *J. Intell. Mater. Syst. Struct.*, 109, 253, 2014.
131. Chen, H., Zhang, X., Liu, J., Tan, C., Compressed Air Energy Storage, In *Energy Storage – Technologies and Applications*, ed. Ahmed F. Zobaa, 2013.
132. Deane, J.P., Gallachóir, B.Ó., Mckeogh, E., Techno-economic review of existing and new pumped hydro energy storage plant. *Renew. Sustain. Energy Rev.*, 14, 1293, 2010.
133. Aramendia, I., Gamiz, U., Vicente, A., Zulueta, E., Guede, J.M., Vanadium redox flow batteries: A review oriented to fluid-dynamic optimization. *Energies*, 14, 176, 2020.
134. Aaron, D., Liu, Q., Tang, Z., Grim, G., Papandrew, A., Turhan, A., Zawodzinski, T., Mench, M., Dramatic performance gains in vanadium redox flow batteries through modified cell architecture. *J. Power Sources*, 206, 450, 2012.
135. Pezeshki, A.M., Clement, J.T., Veith, G.M., Zawodzinski, T.A., Mench, M.M., High performance electrodes in vanadium redox flow batteries through oxygen-enriched thermal activation. *J. Power Sources*, 294, 333, 2015.
136. Houser, J., Pezeshki, A., Clement, J.T., Aaron, D., Mench, M.M., Architecture for improved mass transport and system performance in redox flow batteries. *J. Power Sources*, 351, 96, 2017.
137. He, Z., Li, M., Li, Y., Li, C., Yi, Z., Zhu, J., Dai, L., Meng, W., Zhou, H., Wang, L., ZrO_2 nanoparticle embedded carbon nanofibers by electrospinning technique as advanced negative electrode materials for vanadium redox flow battery. *Electrochim. Acta*, 309, 166, 2019.
138. Gurieff, N., Timchenko, V., Menictas, C., Variable porous electrode compression for redox flow battery systems. *Batteries*, 4, 53, 2018.
139. Kim, K.J., Lee, S.-W., Yim, T., Kim, J.-G., Choi, J.W., Kim, J.H., Park, M.-S., Kim, Y.-J., A new strategy for integrating abundant oxygen functional groups into carbon felt electrode for vanadium redox flow batteries. *Sci. Rep.*, 4, 1, 2014.
140. Yue, L., Li, W., Sun, F., Zhao, L., Xing, L., Highly hydroxylated carbon fibres as electrode materials of all-vanadium redox flow battery. *Carbon*, 48, 3079, 2010.
141. Zhang, X., Wu, Q., Lv, Y., Li, Y., Zhou, X., Binder-free carbon nano-network wrapped carbon felt with optimized heteroatom doping for vanadium redox flow batteries. *J. Mater. Chem. A.*, 7, 25132, 2019.
142. Kazacos, M., Cheng, M., Kazacos, M., Vanadium redox cell electrolyte optimization studies. *J. Appl. Electrochem.*, 20, 463, 1990.
143. Zhang, J., Li, L., Nie, Z., Chen, B., Vijayakumar, M., Kim, S., Wang, W., Schwenzer, B., Liu, J., Yang, Z.., Effects of additives on the stability of electrolytes for all-vanadium redox flow batteries. *J. Appl. Electrochem.*, 41, 1215, 2011.

144. Islam, R., Nolen, C., Jeong, K., Effects of sulfuric acid concentration on volume transfer across ion-exchange membrane in a single-cell vanadium redox flow battery. in *ASME International Mechanical Engineering Congress and Exposition*, American Society of Mechanical Engineers, 2017.
145. Moore, M.A., A base case design and capital cost analysis of an all vanadium redox-flow battery, Master's Thesis, University of Tennessee, 2013. https://trace.tennessee.edu/utk_gradthes/2441/

7

Vanadium-Based Redox Flow Batteries

Anurag Tiwari, Shishir Kumar Singh, Dipika Meghnani, Raghvendra Mishra and Rajendra Kumar Singh*

Ionic Liquid and Solid-State Ionics Lab, Department of Physics, Institute of Science, Banaras Hindu University, Varanasi, India

Abstract

Climate change and global warming are major environmental challenges due to fossil fuel usage worldwide. Effective renewable energy storage technologies are vitally needed for sustainable ecology. Redox flow batteries (RFBs) are a kind of energy storage device that is attracting increased interest because of their cost-effective scaling strategy, ease of component replacement, and extended cycle life. One of the latest RFB technologies being researched to efficiently store renewable energy is vanadium redox flow batteries (VRFBs). Uninterruptible power supplies (UPS), load balancing, power conversion technology, and electric cars all make extensive use of VRFBs. The present need for and status of renewable energy generation, the basic ideas behind VRFBs, their operation, and their technological limitations have all been thoroughly covered in this chapter. Each component's operating principles are noted. The applications of the limiting factors of certain components and past/current research to solve these limits are addressed.

Keywords: Vanadium redox flow batteries, redox reaction, catholyte, anolyte, nanofiltration

7.1 Introduction

Due to the widespread use of conventional fossil fuels in contemporary society, climate change and global warming have emerged as two of the most serious and damaging environmental problems. Additionally, as traditional (hydrocarbon-based) fuels gradually run out of natural resources, there is

**Corresponding author*: rksingh_17@rediffmail.com

Inamuddin and Tariq Altalhi (eds.) Advanced Redox Flow Technology, (161–176) © 2024 Scrivener Publishing LLC

a growing need for alternate energy sources [1, 2]. Owing to their minimal negative effects on the environment, renewable energy sources have drawn considerable attention. Today, a variety of energy storage devices are used to meet the demands of contemporary civilization. Hydroelectric power plants (HEP) and thermoelectric power plants (TEP) are the two that predominate. However, they can only be constructed in certain geographical locations, such as huge water zones for HEPs [3–7]. Renewable energy with high energy and power density is gaining attention as a potential solution to the problem of providing a reliable power source that is not tied to any one location and has a minimal ecological impact. Chemical power sources (CPSs), which store electricity as the chemical potential of redox-active components, have emerged as a practical storage technique in recent decades [8, 9]. Lead-acid battery technology is well liked in the CPS due to its inexpensive investment costs as compared to other types, but their deployment was constrained by their short cycle life and capacity. Other batteries, such as lithium-ion, nickel-metal hydride, nickel-cadmium, and sodium-ion batteries offer higher energy densities and cycle lives, but they also come with higher costs [10, 11]. In the replacement of fossil fuels, one of the most important requirements for a renewable energy source is to provide a low-cost alternative for stored energy. This is why, unlike other electrochemical energy storage devices, the emphasis is on durable, low-cost storage devices rather than compact and simple systems or high power and energy densities [12]. Redox flow batteries (RFBs) are gaining popularity due to their independent modular design, scaling of electrochemical power and energy, simplicity of replacing broken parts, and affordability of the storage device [1, 13]. In this chapter, the fundamentals of redox flow batteries, their operating concepts, and several types of redox flow batteries are covered. In the later part of the chapter, vanadium redox flow batteries are discussed, along with some of their uses.

7.2 Redox Flow Batteries (RFBs)

Because of their unique electrochemical technology, redox flow batteries (RFBs) are extremely important for grid-scale stationary energy storage among CPS. Redox-active electrolytes in tanks are oxidized and reduced in membrane-separated electrode portions of the membrane electrode assembly to generate electrical energy. The notion of RFBs was initially addressed in a patent, in which the electrolyte consists of vanadium salts in addition to other metal salts (Cr, Co, Ce, Fe, Mn, Ti) [14]. As an energy buffer for wind turbines, solar energy plants, and tidal power plants, the inventor of that patent suggested the notion of storing electric power in liquids via redox reactions of dissolved redox couples at electrodes.

To combat issues with energy density, diffusion, and efficiency, the RFBs' fundamental requirement is "a single component, which may exist in three or more oxidation states and utilized as an electrochemically active agent in solution." On this basis, Chromium-redox flow battery (Cr-RFB), Cr^{3+}/Cr^{2+} // Cr^{6+}/ Cr^{3+} and Vanadium-redox flow battery (VRFB), V^{3+}/V^{2+} // V^{5+}/V^{4+}, etc., demonstrated great potential [13]. RFBs offer flexibility and the potential for grid-scale storage, independent scaling of energy and power, long durability, and little environmental effect because the redox active components are stored in the external tanks and pumped in when needed [15].

7.2.1 The General Structure of RFBs

RFBs are electrochemical cells made up of two containers, each containing half of the cell's electrolyte, which is pumped across a membrane to the correct side. As depicted in Figure 7.1, a membrane separates the catholyte and the anolyte (the two liquid electrolytes, by convention). This membrane allows ions (typically H^+ and OH^- in aqueous media) to pass through. Conventionally, the positive electrode has termed the cathode and the negative electrode is termed the anode in electrochemical devices. Single RFB cell or networks of RFB cell is known as RFB. The redox-flow battery is a system that consists of cells, fluid flow technologies, pumps, electrolyte storage, battery management systems (BMS), etc. [13, 16].

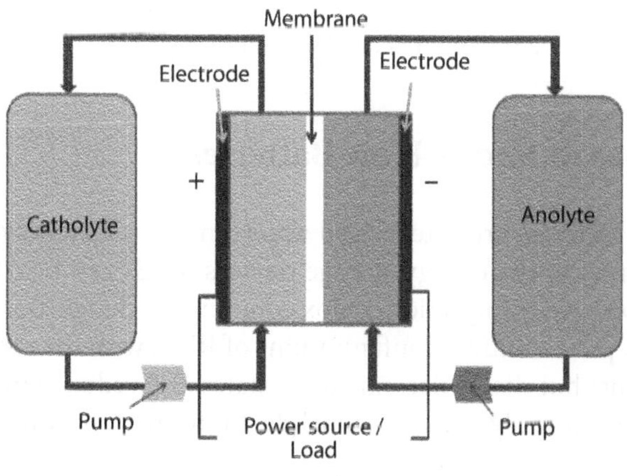

Figure 7.1 Basic structure of redox flow battery.

7.2.2 Working of Redox Flow Batteries

These devices produce electrical power through the oxidation and reduction of separately stored redox-active electrolytes, which are stored in the tanks and pumped through the membrane electrode assembly's membrane-separated electrode compartments [13]. In the discharge mode, the working principle of redox-flow batteries can be broadly characterized as a chemical reaction between two redox couples that results from the combination of two related half-reactions as follows:

Anode:

$$A^{x} \leftrightarrow A^{x+y} + ye^{-} \qquad \varphi_A \tag{7.1}$$

Cathode:

$$B^{z} + ye^{-} \leftrightarrow B^{z-y} \qquad \varphi_B \tag{7.2}$$

Cell reaction

$$A^{x} + B^{z} \leftrightarrow A^{x+y} + B^{z-y} \qquad \varphi_{cell} = \varphi_B - \varphi_A \tag{7.3}$$

The redox couple A^{x} / A^{x+y} is used for positive electrode reaction and B^{z-y} / B^{z} is used for negative electrode reaction. The cell voltage of RFBs is defined as the difference between the redox pairs, mentioned above. The release of ion and electron(s) from electrolyte during half-reaction at one electrode while at another electrode, recombination reaction into electrolyte occurs. The migration of ions takes place through the electronically insulating membrane, thus the electron is forced to move through the external circuit which provides electrical energy [17].

7.3 Types of Redox Flow Batteries

Redox flow batteries are categorized according to the catholyte and anolyte that consist of the energy storage devices. RFBs can be configured to operate in ion carrier migration, redox coupling, or charge-transfer modes. Figure 7.2 depicts multiple configurations of RFBs with the same elements or constituents but dissimilar oxidation states and redox active species in the catholyte and anolyte utilized in RFBs. In this part, we will cover briefly

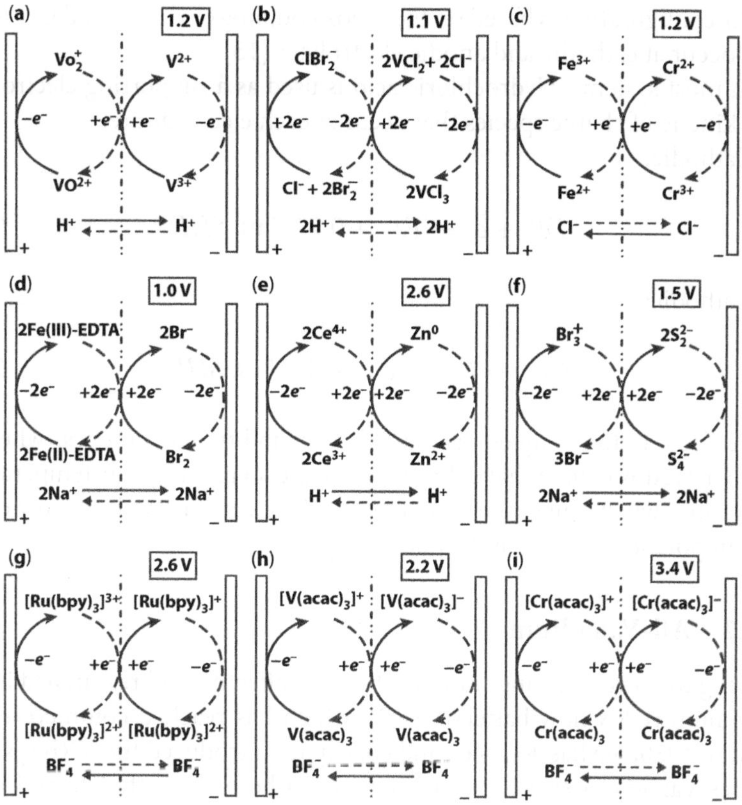

Figure 7.2 Schematic representation of charge migration in different RFB systems (the values provide the cell voltage of the redox couple). (a) All vanadium, (b) vanadium/bromine, (c) iron/chromium, (d) Fe-EDTA/bromine, (e) zinc/ cerium, (f) bromine/polysulphide, (g) nonaqueous ruthenium/bipyridine, (h) nonaqueous vanadium/acetylacetonate, (i) nonaqueous chromium/acetylacetonate. Reproduced with permission from Reference [DOI 10.1007/s10800-011-0348-2]. Copyright 2011 under Creative common CC by License, Springer Nature.

some of the configurations of RFBs, as well as their advantages, disadvantages, and applications.

7.3.1 Iron/Chromium

This was the earliest redox flow battery, created by NASA in 1970. It is utilized in photovoltaic array applications with a power of 1 kW/13 kWh. The

Fe/Cr configuration is based on the redox couplings Fe^{3+}/Fe^{2+} and Cr^{3+}/Cr^{2+} that occur at cathodic and anodic electrolytes [15].

In most systems, Hydrochloric acid is used as a supporting electrolyte. The kinetic of charge species that occur at each electrode are:

Cathodic:

$$Fe^{3+} + e^{-} \leftrightarrow Fe^{3+} + 0.77V \ vs. \ SHE \tag{7.4}$$

Cathodic:

$$Cr^{2+} \leftrightarrow Cr^{3+} - e^{-} - 0.41V \ vs. \ SHE \tag{7.5}$$

Fe^{3+}/Fe^{2+} redox couples have good reversibility and kinetics, whereas Cr^{3+}/Cr^{2+} redox couples have slow kinetics and so require high-temperature operation, which raises the system's cost. This system also has a low open circuit potential (1.0 V) and a poor energy density.

7.3.2 All-Vanadium

Among the several configurations, the most effective and researched is the vanadium redox flow battery (VRB), which has reached commercialization. In 1980, Skyllas-Kazacos and colleagues introduced the VRB system. It uses Vanadium entirely as an element in both the catholyte and anolyte, which are dissolved in an aqueous sulphuric acid solution, providing a cell voltage of 1.26 V. In all VRB systems, the V^{2+}/V^{3+} and V^{4+}/V^{5+} redox couplings occur at the negative and positive electrodes, respectively [18, 19]. At the positive and negative electrodes, the following electrochemical half-reactions occur:

$$\text{Positive Electrode: } VO^{2+} + H_2O \leftrightarrow VO_2^{+} + 2H^{+} + e^{-} + 1.0 \ V \ vs. \ SHE \tag{7.6}$$

$$\text{Negative Electrode: } V^{2+} \leftrightarrow V^{3+} + e^{-} - 0.26V \ vs. \ SHE \tag{7.7}$$

In this case, the migration of hydrogen ions, $2H^{+}$, through the membrane separator, maintains the current. A VRB cell's typical open circuit voltage is 1.26 V, but with Nernst's factor correction, its true OCV is closer to 1.4 V. Compared to competitors like lithium-ion batteries, VRBs have a current density of 50–80 mA/cm^2 and a power density of 0.1 W/cm^2, which is substantially less. Vanadium solubility is in the electrolyte stream and precipitation formations are connected to the decreased power density.

7.3.3 Vanadium/Bromine

For better VRB cell output, halide ions are utilized because they increase vanadium solubility. The catholyte and anolyte of the vanadium-bromine battery, which serves as the supporting electrolytes, undergo VBr_3/VBr_2 and Br-/ClBr- redox reactions, respectively, when dissolved in the electrolytes of HBr and HCl [20, 21]. The following redox processes take place at the positive and negative electrodes:

$$\text{Positive Electrode: } Br^- + Cl^- \leftrightarrow ClBr_2^- + 2e^- + 1.09\ V\ vs.SHE \quad (7.8)$$

$$\text{Negative Electrode: } 2VBr_3 + 2e^- \leftrightarrow 2VBr_3 + 2Br^- - 0.26V\ vs.SHE \quad (7.9)$$

Vanadium's increased solubility raises the energy density of the current configuration; yet this system has some drawbacks, including the emission of bromine vapor.

7.3.4 Bromine/Polysulfide

In the bromine/polysulfide system, the catholyte and anolyte are sodium bromide and sodium polysulfide, respectively. The following electrochemical reactions occur at positive and negative electrodes:

$$\text{Positive Electrode: } 3Br^- \leftrightarrow Br_3^- + 2e^- + 1.09V\ vs.\ SHE \quad (7.10)$$

$$\text{Negative Electrode: } 2S_2^{2-} \leftrightarrow S_4^{2-} + 2e^- - 0.265V\ vs.\ SHE \quad (7.11)$$

The sulfur in the solution is translocated between polysulfide and sulfide at the negative electrode, while three bromide ions form the tribromide ion at the positive electrode [22, 23].

7.4 Vanadium Redox Flow Battery (VRFB)

The vanadium redox flow battery (VRFB) seems to be the most reliable among the numerous redox flow batteries because it has good stability, a big capacity, a long cycle, and safe operation. Skyllas-Kazacos *et al.* first presented the VRFB concept in the 1980s [24, 25]. In the two liquid electrolytes, the same material (vanadium) is used; therefore, cross-contamination problems and hence capacity loss due to this are not calamitous as in the dissimilar liquid in each electrolyte. Cross-contamination issues and capacity loss, as a result, are not catastrophic as in the different liquids in each electrolyte because the same substance (vanadium) is employed in the two liquid electrolytes.

A VRFB only needs to be recharged to prevent electrolyte cross-contamination, which results in extremely little electrolyte degradation and extended cycle life [24]. Redox flow batteries outperform other traditional battery types in terms of system flexibility, safety considerations, quick reaction, and expenditure for grid-scale applications, and they are anticipated to gain more commercial traction as technology advances in the coming years. Therefore, it is essential to comprehend the technological foundations and redox chemistry of the flow batteries, which govern their technology and market dominance [26]. Figure 7.3 schematically depicts the vanadium redox flow batteries' structure and charge-discharge reactions. When discharging, the reaction can be expressed as

At positive electrode:

$$V^{2+} \leftrightarrow V^{3+} + e^- \qquad (7.12)$$

At negative electrode:

$$VO^{2+} + 2H^+ + e^- \leftrightarrow VO_2^+ + H_2O \qquad (7.13)$$

Overall cell reaction:

$$V^{2+} + VO^{2+} + 2H^+ \leftrightarrow V^{3+} + VO_2^+ + H_2O \qquad (7.14)$$

The cell potential for the all-vanadium RFBs is 1.26 V. For the given concentrations, particular temperature, and fixed pH value of vanadium species, the cell voltage can be calculated based on the Nernst equation:

$$E = 1{:}26V - RT/F \ln([VO^{2+}].[V^{3+}]/[VO_2^{+}].[H^{+}].[V^{2+}]) \quad (7.15)$$

Where, F, T, and R are the Faraday constant, absolute temperature and universal gas constant, respectively.

7.4.1 Working Principle of Vanadium Redox Flow Battery

The vanadium redox flow battery (VRFB) is the most promising, well-studied RFB generation. The storage of electrolytes is the primary distinction between RFBs and other electrochemical batteries. The electrolyte is kept outside the battery's core in a flow battery. Vanadium ions are kept in tanks of VRFBs in four distinct valance states VO^{2+}, V^{3+}, V^{2+} (which would be V^{4+}), and VO_2^{+} (which would be V^{5+}) (see Figure 7.3). Overall, the electrolyte and electrode that are associated with V^{4+} or V^{5+} are on the positive side, while those on the negative side are the electrolyte and electrode.

Electrolytes are absorbed by each half-cell in the RFB. They are then pumped into a storage reservoir for recirculation [27]. Every half-cell of the RFB is a combination of a bipolar plate and an electrode; each half-cell is divided by a membrane that allows ion diffusion while avoiding electrolytes from being contaminated [25]. This configuration features a single cell that shares form the stack of bipolar plates. Eqs. (7.12) and (7.13) show the reactions on the cathode and anode electrodes in a half-cell.

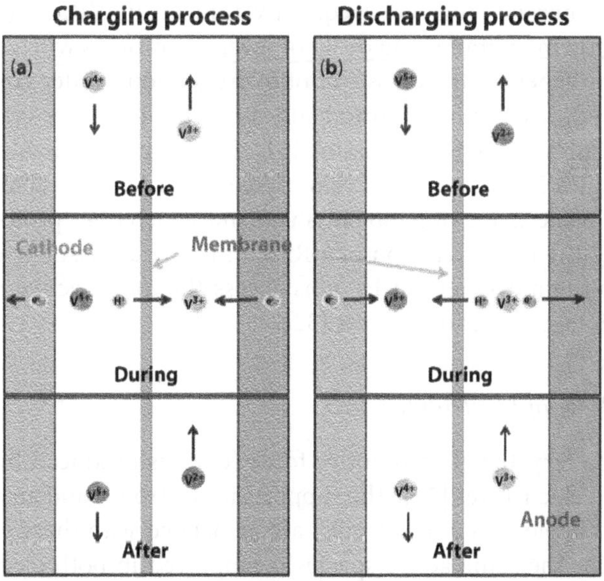

Figure 7.3 General mechanism of VRFBs during (a) charging (b) discharging process.

In the discharge stage, the positive and negative electrolytes comprise only V^{3+} and V^{4+} (VO^{2+}). The positive electrolyte is oxidized whereas the negative electrolyte is decreased during charging [28]. During this time, electrons travel from the positive to the negative side of the bipolar plate, while hydrogen ions diffuse on the negative side across the membrane [29].

7.4.2 Role of Different Components in VRFBs

7.4.2.1 Role of Membrane

The membrane of VRFBs plays a crucial role in facilitating proton crossover and preventing electrolytes from getting contaminated on both the positive and negative sides. The performance of the flow batteries is significantly impacted by the membrane's efficiency after numerous cycles. The fundamental issue for the membrane in the VRFBs is to maintain both chemical stability and proton conductivity at the same time. A larger pore size not only improves proton conductivity but also encourages vanadium ion crossover, which can reduce coulombic efficiency. Ion flow is made easier by thinner membranes since they have lower resistance, but functioning also requires that mechanical strength be maintained. To attain great efficiency, the membrane's pore size and thickness must be balanced [25, 30].

Numerous membranes have been designed in consideration of the issues mentioned. The Nanofiltration Membrane and Ion Exchange Membrane are two membranes that are frequently utilized in VRFBs (IEM). The nanofiltration membrane is made up of several tiny pores with certain pore sizes that only permit the crossover of hydrogen ions under conditions of enough pressure difference. Proton and vanadium ions have very different ionic radii, preventing their crossing [31]. Ion Exchange Membrane refers to the complex 3D networks of cross-linked linear polymer chains that are generated in the shape of a tube or sheet and which offer proton conduction during the redox reaction of VRFB. This characteristic is utilized by VRFBs for the movement of H^+ ions across the ion-selective membrane and to maintain the redox process [32].

7.4.2.2 Role of Electrolyte

In a flow battery, the electromotive force (emf) is produced by the reaction of both electrolytes [33]. The supporting electrolyte and an electro-active redox couple (salt and solvent) are associated with these electrolytes (Figure 7.3). The same active species is employed in both electrolytes in the VRFBs, resulting in electrolyte cross-contamination, which is the primary factor in the reduced capacity loss. Depending on the concentration, VRFBs can normally store electrolytes at a rate of 20 to 30 Wh/L. Based on concentration, VRFBs can typically store 20–30 Wh/L of catholyte and anolyte [34]. The last few decades have seen the consideration of a variety of supporting electrolytes for several vanadium compounds, including vanadium pentoxide (V_2O_5), vanadium trichloride (VCl_3), and vanadyl sulfate ($VOSO_4$), along with hydrochloric acid (HCl), sulfuric acid (H_2SO_4), and sodium hydroxide (NaOH) [35]. Because V_2O_5 has very low solubility, it was also removed to prevent the generation of chlorine gas along with VCl_3 and HCl. At first, $VOSO_4$ and H_2SO_4 were chosen. To create electrolytes using V_2O_5, however, certain techniques were established. Many scientists are currently working to employ V_2O_5 to lower the price of electrolytes [36, 37].

7.4.2.3 Role of Electrode

In the flow battery, the electrode facilitates the redox (oxidation and reduction) process. Electrodes of VRFBs are designed porous to provide larger sites for the reaction and it also acts as a catalyst for redox reactions [38]. Generally, VRFBs have two electrodes: positive (facilitating eq. 7.12) and negative (facilitating eq. 7.13). Oxidation-reduction occurring at electrodes must be reversible and the constituents component must be electrolyte-soluble. Stable states must be maintained through the reduction and oxidation reactions to enhance the redox potential. Materials of the electrode are chosen such that it acts as a good catalyst for redox reactions and promotes the flow of charged electrolyte. Since the supporting electrolytes have a highly acidic nature, electrode materials need to be chosen such that they must be corrosion resistive. Generally in the VRFBs, sulfuric acid-based electrolytes are used. Thus carbon and graphite-based electrodes are commonly used due to their anti-corrosive nature and good electrical conductivity [39, 40].

7.5 Applications of Vanadium Redox Flow Batteries (VRFBs)

In addition to more conventional battery types, VRFBs are already in use for a variety of immobile power storage applications. However, due to their specificity, they are often used in low-power/energy applications. VRFBs' benefit, which is not reliant on stored energy or rated power in many applications, will eventually overcome the disadvantages of VRFBs, for example, lower energy density and more complexity. Uninterruptible power supplies (UPS), peak shaving, combination with renewable energy, and load leveling (wind and solar power output) are some popular applications for VRFBs. A load-leveling chemical power source (CPSs) is utilized to provide consistent power usage throughout time. To reduce expenses, it stores power during periods of low energy prices and subsequently provides electrical energy during times of premium price [41]. In several industries, there are rare but notable cases of very high-power utilization during relatively brief periods. During peak power events, it is common to practice minimizing the demand for immediate power supply from the electrical grid to meet a lower contractual peak power and avoid this additional cost [42]. For instance, Sumitomo Electric Industries Ltd. installed a 1.5 MW/1.5 MWh G1 system at Tottori Sanyo Electric in Japan in 2001 for a peak shaving application [43]. A 200 kW/800 kWh G1 system at the Kashima-Kita Electric Power was installed by Mitsubishi Chemicals for load leveling in Japan in 1996 as an example of this usage [41]. Future uses for peak shaving of flow batteries may include rapid charging of electric cars at current gas stations. Although flow batteries do not yet have the energy density required in the electric vehicle (EV) installation, they might be used to charge EVs during periods of high electrical power demand, transferring the usage of the electrical energy to off-peak times when the battery was charged by low-contract electricity from the grid, which may be performed by storing VRFB electrolyte in deactivated gas station tanks. This method would have the advantage of laying the groundwork for the future supply of flow battery-equipped vehicles, if they prove to be economically viable, along with the micro-grid benefits of this arrangement (integration of local electricity production, storage of electricity for auto consumption throughout brief blackouts). Additionally, the progressive adaption of deactivated gas tanks for flow battery electrolyte stockpiling would allow gas stations to seamlessly migrate from the fossil-fuel-based economy to the electric mobility paradigm following demand change. Even if, in the future, flow batteries are solely employed to shorten EV charging periods during peak demand, this theory would still be applicable. In a separate, ongoing journal article, the experts assess this strategy's technical, energy, and economical feasibility. VRFBs have a long cycle life, and uninterruptible power supplies (UPS) technologies may offer an uninterrupted electricity supply and/or a clean operating voltage with minimal total harmonic distortion (THD) [44]. A 500 kW/90 kWh G1 system is installed by Sumitomo Electric Industries Ltd. in a factory in Japan in 2003 for peak shaving and UPS purposes [43]. VRFBs may also be used in conjunction with renewable energy sources. In particular, the instability of the grid is harmed by the fact that wind and solar power are often intermittent. An energy buffer is usually used to store extra energy for subsequent release during peak demand because their power generation frequently does not match their power demand. This is particularly helpful in the case of solar energy and wind turbines, which characteristically reach their peak output at midday and night, VRFBs are used to store that energy in the form of CPSs and deliver when electrical energy is required. While many hydropower facilities can pump energy backward and store it, wind turbines are commonly sited in remote areas and require confined storage to maximize output [45]. The electrical energy storage in the form of VRFB technology is especially useful for structures without an electrical grid connection, and exterior inhabited regions, which may employ a wind power, solar, or microhydraulic system to create electrical power and store it in the form of CPSs for later consumption when needed. For all the aforementioned cases, the VRFBs appear to be a suitable solution. In reality, VRFBs are being studied for use in solar and wind power generation, for grid-level application [7, 46–48].

7.6 Summary

In power grid nodes, where irregular electrical energy output and consumption must be balanced, redox flow batteries (RFBs) are reliable energy storage technologies. Among all the redox flow batteries, vanadium-based RFBs (VRFBs) have been widely studied and implemented as high capacity, long cycle life, and negligible self-discharge energy storage systems. In the VRFBs, electrodes and membranes play a crucial role in the production and storage of electrical power in the form of chemical potential. Electrodes provide sites for redox reactions while the membrane provides a path for proton exchange and restricts electronic flow, forcing electron movement through the external circuit. The system may be enhanced with better electrolyte compositions and more efficient membrane, electrode, and bipolar plate materials. For VRFBs to be used in commercial applications and to maintain high efficiency, the stack's resistive and parasitic losses must be reduced.

References

1. Petrov, M.M., Modestov, A.D., Konev, D.V., Antipov, A.E., Redox flow batteries : Role in modern electric power industry and comparative characteristics of the main types. *Russ. Chem. Rev.*, 90, 677, 2021.

2. Kumar, S., Dutta, D., Kumar, R., Enhanced structural and cycling stability of Li_2CuO_2-coated polymer electrolyte for lithium polymer batteries. *Electrochim. Acta,* **343,** 136122, 2020.
3. Berrada, A., Loudiyi, K., *Gravity Energy Storage*, pp. 1–171, Elsevier, 2019.
4. Li, C., Zhang, B., Liu, Q., N-eicosane/expanded graphite as composite phase change materials for electro-driven thermal energy storage. *J. Energy Storage,* **29,** 101339, 2020.
5. Rosa, L.P., Santos, M.A., Gesteira, C., Xavier, A.E., A model for the data extrapolation of greenhouse gas emissions in the Brazilian hydroelectric system. *Environ. Res. Lett.*, **11**, 064012, 2016.
6. Satpute, S.A., Shinde, S.S., Dragon fruit : A review of health coverage and nutrients. *Intl. J. Res. Publ. Rev.*, **3,** 2693–702, 2022.
7. Divya, K.C., Østergaard, J., Battery energy storage technology for power systems - An overview. *Electr. Power Syst. Res.,* **79,** 511–520, 2009.
8. Osman, M., Khan, M.N., Zaabout, A., Cloete, S., Amini, S., Review of pressurized chemical looping processes for power generation and chemical production with integrated CO_2 capture. *Fuel Process. Technol.,* **214,** 106684, 2021.
9. Mishra, R., Singh, S.K., Gupta, H., Tiwari, R.K., Meghnani, D., Patel, A., Tiwari, A., Tiwari, V.K., Singh, R.K., Polar β - Phase PVdF-HFP-Based Freestanding and Flexible Gel Polymer Electrolyte for Better Cycling Stability in a Na Battery. *Energy Fuels*, 35, 15153-15165, 2021.
10. Meghnani, D., Gupta, H., Singh, S.K., Srivastava, N., Mishra, R., Tiwari, R.K., Patel, A., Tiwari, A., Singh, R.K., Enhanced cyclic stability of $LiNi_{0.815}Co_{0.15}Al_{0.035}O_2$ cathodes by surface modification with $BiPO_4$ for applications in rechargeable lithium polymer batteries. *ChemElectroChem.,* **8,** 2867–2880, 2021.
11. Guo, S., Feng, Y., Wang, L., Jiang, Y., Yu, Y., Hu, X., Architectural engineering achieves high-performance alloying anodes for lithium and sodium ion batteries. *Small,* **17,** 1–26, 2021.
12. Budischak, C., Sewell, D., Thomson, H., MacH, L., Veron, D.E., Kempton, W., Cost-minimized combinations of wind power, solar power and electrochemical storage, powering the grid up to 99.9% of the time. *J. Power Sources,* **225,** 60–74, 2013.
13. Noack, J., Roznyatovskaya, N., Herr, T., Fischer, P., The chemistry of redox-flow batteries. *Angew. Chemie - Int. Ed.,* **54,** 9776–9809, 2015.
14. Thaller, L.H., U.S. Pat. 3,996, 1976.
15. Weber, A.Z., Mench, M.M., Meyers, J.P., Ross, P.N., Gostick, J.T., Liu, Q., Redox flow batteries: A review. *J. Appl. Electrochem.,* **41,** 1137–1164, 2011.
16. Aramendia, I., Gamiz, U., San-vicente, A., Zulueta, E., Guede, J.M., Vanadium redox flow batteries: A review oriented to fluid-dynamic optimization. *Energies* **14,** 176, 2021.
17. Hagedorn, N.H., Thaller, LH., Design flexibility of redox flow systems. *NASA Tech Memo,* 1982.

18. Rahman, F., Kazacos, M., Vanadium redox battery: Positive half-cell electrolyte studies. *J. Power Sources,* **189,** 1212–1219, 2009.
19. Vijayakumar, M., Burton, S.D., Huang, C., Li, L., Yang, Z., Graff, G.L., Liu, J., Hu, J., Maria, S.K., Nuclear magnetic resonance studies on vanadium(IV) electrolyte solutions for vanadium redox flow battery. *J. Power Sources,* **195,** 7709–7717, 2010.
20. Kazacos, M., Novel vanadium chloride/polyhalide redox flow battery. *J. Power Sources,* **124,** 299–302, 2003.
21. Vafiadis, H., Kazacos, M., Evaluation of membranes for the novel vanadium bromide redox flow cell. *J. Memb. Sci.,* **279,** 394–402, 2006.
22. Kalafut, A.E.J., United States Patent (19), 1984.
23. Lessner, P.M., McLarnon, F.R., Winnick, J., Cairns, E.J., Aqueous polysulphide flow-through electrodes: Effects of electrocatalyst and electrolyte composition on performance. *J. Appl. Electrochem.,* **22,** 927–934, 1992.
24. Kazacos, M., All-Vanadium Redox Battery United States Pat., 169, 070513, 1988.
25. Prifti, H., Parasuraman, A., Winardi, S., Lim, T.M., Kazacos, M., Membranes for redox flow battery applications. *Membranes (Basel),* **2,** 275–306, 2012.
26. Chen, R., Kim, S., Chang, Z.,. Redox flow batteries: Fundamentals and applications, in: *Redox: Principles and Advance Applications,* M.A.A. Khalid (Ed.), 2017.
27. Bartolozzi, M., Development of redox flow batteries. A historical bibliography. *J. Power Sources,* **27,** 219–234, 1989.
28. Gundlapalli, R., Kumar, S., Jayanti, S., Stack design considerations for vanadium redox flow battery. *Ina. Lett.,* **3,** 149–157, 2018.
29. de León, C., Ferrer, A., García, J., Szánto, D.A., Walsh, F.C., Redox flow cells for energy conversion. *J. Power Sources,* **160,** 716–32, 2006.
30. Schwenzer, B., Zhang, J., Kim, S., Li, L., Liu, J., Yang, Z., Membrane development for vanadium redox flow batteries. *ChemSusChem.,* **4,** 1388–406, 2011.
31. Zhang, H., Zhang, H., Li, X., Mai, Z., Zhang, J., Nanofiltration (NF) membranes: The next generation separators for all vanadium redox flow batteries (VRBs)? *Energy Environ. Sci.,* **4,** 1676–1679, 2011.
32. Xu, T., Ion exchange membranes: State of their development and perspective. *J. Memb. Sci.,* **263,** 1–29, 2005.
33. Blanc, C., Rufer, A., Understanding the Vanadium Redox Flow Batteries, in: *Paths to Sustainable Energy,* J. Nathwani, A. Ng (Ed.), IntechOpen, 2010.
34. Zhang, G.P., Gao, G.J., Wang, X.Q., Lv, G.H., Zhou, L., Chen, H., Pang, H., Yang, S.Z., Influence of pulsed substrate bias on the structure and properties of Ti-Al-N films deposited by cathodic vacuum arc. *Appl. Surf. Sci.,* **258,** 7274–7279, 2012.
35. Yang, Z., Tong, L., Tabor, D.P., Beh, E.S., Goulet, M.A., De Porcellinis, D., Guzik, A., Gordon, R.G., Aziz, M.J., Alkaline Benzoquinone aqueous flow battery for large-scale storage of electrical energy. *Adv. Energy Mater.,* **8,** 1–9, 2018.

36. Choi, C., Kim, S., Kim, R., Choi, Y., Kim, S., Jung, H.Y., Yang, J.H., Kim, H.T., A review of vanadium electrolytes for vanadium redox flow batteries. *Renew. Sustain. Energy Rev.*, **69,** 263–74, 2017.
37. Li, W., Zaffou, R., Sholvin, C.C., Perry, M.L., She, Y., Vanadium redox-flow-battery electrolyte preparation with reducing agents. *ECS Trans.*, **53,** 93–99, 2013.
38. Couper, A.M., Pletcher, D., Walsh, F.C., Electrode materials for electrosynthesis. *Chem. Rev.*, **90,** 837–865, 1990.
39. Wang, W., Luo, Q., Li, B., Wei, X., Li, L., Yang, Z., Recent progress in redox flow battery research and development *Adv. Funct. Mater.*, **23,** 970–986, 2013.
40. Flox, C., Skoumal, M., Garcia, J., Andreu, T., Morante, J.R., Strategies for enhancing electrochemical activity of carbon-based electrodes for all-vanadium redox flow batteries. *Appl. Energy,* **109,** 344–351, 2013.
41. Wagner, R., Large lead/acid batteries for frequency regulation, load levelling and solar power applications. *J. Power Sources,* **67,** 163–72, 1997.
42. Johnson, M.P., Noy, A., Liu, O., Feng, Y., Energy peak shaving with local storage. *Sustain. Comput. Informatics Syst.*, **1,** 177–88, 2011.
43. Shigematsu, T., Kumamoto, T., Deguchi, H., Hara, T., Applications of a vanadium redox-flow battery to maintain power quality. *Proc. IEEE Power Eng. Soc. Transm. Distrib. Conf.*, **2,** 1065–70, 2002.
44. Zhan, Y., Wang, H., Zhu, J., Modelling and control of hybrid UPS system with backup PEM fuel cell/battery. *Int. J. Electr. Power Energy Syst.*, **43,** 1322–31, 2012.
45. Sarkar, J., Bhattacharyya, S., Application of graphene and graphene-based materials in clean energy-related devices. Minghui *Arch. Thermodyn.*, **33,** 23–40, 2012.
46. Rydh, C.J., Sandén, B.A., Energy analysis of batteries in photovoltaic systems. Part II: Energy return factors and overall battery efficiencies. *Energy Convers. Manag.* **46,** 1980–2000, 2005.
47. Joerissen, L., Garche, J., Fabjan, C., Tomazic, G., Possible use of vanadium redox-flow batteries for energy storage in small grids and stand-alone photovoltaic systems. *J. Power Sources,* **127,** 98–104, 2004.
48. Li, W., Joós, G., Bélanger, J., Real-time simulation of a wind turbine generator coupled with a battery supercapacitor energy storage system. *IEEE Trans. Ind. Electron.*, **57,** 1137–1145, 2010.

8

System for the Redox Flow Technology

Shrabani Barman[1] and Ujjwal Mandal[2*]

[1]Department of Chemistry, Krishnath College, Mushidabad, India
[2]Department of Chemistry, The University of Burdwan, Golapbag, Burdwan, India

Abstract
Urbanization and rapid population growth have increased daily life energy consumption manyfold. Most of the energy production comes from burning fossil fuels, which has a permanent harmful impact on mankind as well as on the environment. Hence, we need alternative energy sources, and renewable energies like solar and wind have become the best alternatives for the same. Now, the major concern is to store the generated energy and use it on demand. For such purposes, redox flow technology is the best example where power can be stored efficiently. The Redox flow battery (RFB) is one of the greatest achievements of scientists because it can store power and energy separately. RFB has drawn a great deal of attention due to its sustainability, flexibility, independent scalability for storage capacity, the highest recyclability of raw materials, long lifetime, and operational safety.

Keywords: Redox flow battery, electrochemical energy, power, renewable energy, stability, long lifetime

8.1 Introduction

More sophisticated lifestyles increase energy consumption, and further energy consumption increases worldwide due to the growing population and booming industrial revolutions. Most energy sectors to date mainly depend on fossil fuels, which bring the highest amount of carbon emission and are also related to many other pollutions. So, an urgent substitute

**Corresponding author*: umandal@chem.buruniv.ac.in

Inamuddin and Tariq Altalhi (eds.) Advanced Redox Flow Technology, (177–198) © 2024 Scrivener Publishing LLC

is required to reduce the negative impact on the environment [1–3]. Consumption of fossil fuels increases day by day, and reports show that it has roughly doubled since 1980 [4]. Though the type of fossil fuel has been shifted from coal to oil and gas, energy consumption has increased. A chart has been prepared based on fossil fuel (coal, oil, and gas) consumption per year by different countries and it shows that first-world countries like the United States, Australia, Germany, and the UK occupy the top position of the chart. The chart also shows that the total consumption of fossil fuels per annum in the United States is 63130 KWh, which is an incredibly large number. We understand that the use of the huge amount of fossil fuels is slowly and severely damaging our environment and it's not only going to affect our present but also the upcoming future generations. Despite having a clear vision of this dark side of energy consumption from fossil fuels, we would not be able to stop using them. The only way out is to switch to another environment-friendly renewable energy resource like solar energy, or wind energy [5–7].

Renewable energy source is the only option to cut the cost of energy generation and delivery. In the pre-industrial age, solar energy was the only option to fulfill energy demands but with the growing population, it became incapable of fulfilling people's needs, and mankind became prone to use fossil fuels. As we know, every blessing comes with a curse, and when curses become a major stakeholder, then it would be high time to cut them off. Similarly, for us, this is the right time to reduce the consumption of fossil fuels and increase the use of renewable energy. The use of renewable energy gives a clean corridor to mankind to subsidize carbon emissions and other related pollution. A good result has been found based on the use of renewable energy from the decade of 2011-2021. People increased their use from 20% to 28% of the global electricity supply using renewable energy and cut down the use of fossil fuels from 68% to 62% [8, 9]. According to scientists and analysts, by 2050, renewable energy can become our primary source of energy. Till then we need a gap-filling source of energy that will be low-cost, environment-friendly, and safely developed energy storage. In this context, Redox Flow Battery (RFB) would be the best option. Redox flow technology is one of the most recent voltage storage technologies where power and energy are separated [10–12]. The most attractive properties are that the raw materials are fully recyclable, there is a high level of operational reliability and a long life span, and only by increasing the storage tank capability can we increase the power-producing ability [13].

8.2 General Construction of Redox Flow Battery

The construction of a redox flow battery [14–16] is very similar to an electrochemical cell, where chemical changes produce a direct electrical potential. As with electrochemical cells, RFB has two half-cell compartments: i) anode half-cell, and ii) cathode half-cell (Figure 8.1). Two half-cells are separated by an ion-exchange membrane through which only ions can be transferred, not electrolytes. The basic difference between RFB and electrochemical cells is that in RFB, electrolytes are kept in a different storage tank and according to the requirement electrolytes are pumped into reaction vessels, while in an electrochemical cell electroactive materials are stored with the respective electrodes. On the other hand, simply by changing the tank storage capacity we can enhance the power of RFB, which is not possible for the electrochemical cells. As RFB is a rechargeable electrical energy storage device, the raw material is recyclable. During discharge, oxidation in the anode and reduction in the cathode occurs in the RFB system and the reverse process happens during charging.

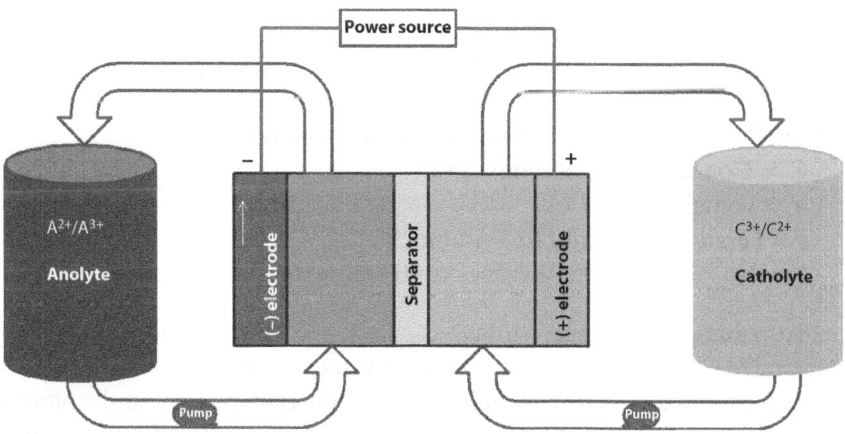

Discharge Reaction:
Cathode: $C^{3+} + e = C^{2+}$ (reduction)
Anode: $A^{2+} = A^{3+}$ e (oxidation)

Charging Reaction:
Cathode: $C^{2+} = C^{3+} + e$ (oxidation)
Anode: $A^{3+} + e = A^{2+}$ (reduction)

Figure 8.1 General formulation of redox flow battery.

Discharge Reaction:

$$\text{Cathode: } C^{3+} + e = C^{2+} \text{ (reduction)}$$

$$\text{Anode: } A^{2+} = A^{3+} + e \text{ (oxidation)}$$

Charging Reaction:

$$\text{Cathode: } C^{2+} = C^{3+} + e \text{ (oxidation)}$$

$$\text{Anode: } A^{3+} + e = A^{2+} \text{ (reduction)}$$

8.3 Energy Capacity

The energy capacity [17] of an RFB is measured by energy density, which is denoted by E (based on active material used as electrolytes). The equation is

$$\text{E (Energy Density)} = (NC_aFV)/n$$

Where N = Total no. of e-s take part in the redox reaction
C_a = Concentration of redox species
V = Voltage
F = Faraday Constant (26.8 $Ah.mol^{-1}$)
N = No. of electrolyte Volumes contributing to redox reactions.

To enhance the energy storage capacity per unit volume, we need to maximize Ca and V, and minimize the n value.

The power generation depends on the current density and voltage. Generally, a flow battery operates at a current density of 50 mA/cm^2. However, the type of active material and ion transfer material used in flow battery construction has a significant role in cell efficiency reduction. The use of a thinner selected membrane can enhance the efficiency to the current density of 80 mA/cm^2 in a 50-kW total output system [18]. Table 8.1 shows the different energy efficiency of batteries for different RFB systems.

Table 8.1 Energy density comparison among different flow battery [18].

Battery name	Energy density (Whl^{-1})	Energy density (Wl^{-1})
Vanadium-Bromine	20-35	50
Vanadium-Vanadium	20-35	60-100
Bromine-Polysulfide	20-35	60
Zinc-Bromine	20-35	40
Zinc-Cerium	20-35	50

8.4 Optimization

To get the optimized efficiency of the cell, we need to consider several other efficiencies too, like voltage efficiency (η_v), charge efficiency (η_c), power efficiency (η_p), and energy efficiency (η_E).

$$\eta_V = (V_{discharge} / V_{charge}), \qquad \eta_c = (Q_{discharge} / Q_{charge}),$$

$$\eta_p = (P_{discharge} / P_{charge}), \qquad \eta_E = (E_{discharge} / E_{charge})$$

8.5 Classification of RFB Based on Active Electrolyte

The conventional flow batteries have two compartments, catholyte and anolyte. They remain separated by a membrane, through which only ions can pass. Based on the active electrolytes, redox flow batteries are divided into two major categories: i) inorganic flow battery, and ii) organic flow battery. Each of the flow batteries has been subdivided into two subcategories, which are membrane and membraneless flow batteries. Each type of flow battery is further divided into full-flow and semi-flow batteries (Figure 8.2).

In an aqueous medium, the most active passing ions are H^+ (349.8 $S.cm^2.mol^{-1}$) and OH^- (198.0 $S.cm^2.mol^{-1}$). If the dissolved active ions on both sides remain the same, there is the least chance of cross-contamination, while the use of two different ions leaves the chance of cross-contamination and may also cause irreversible performance loss.

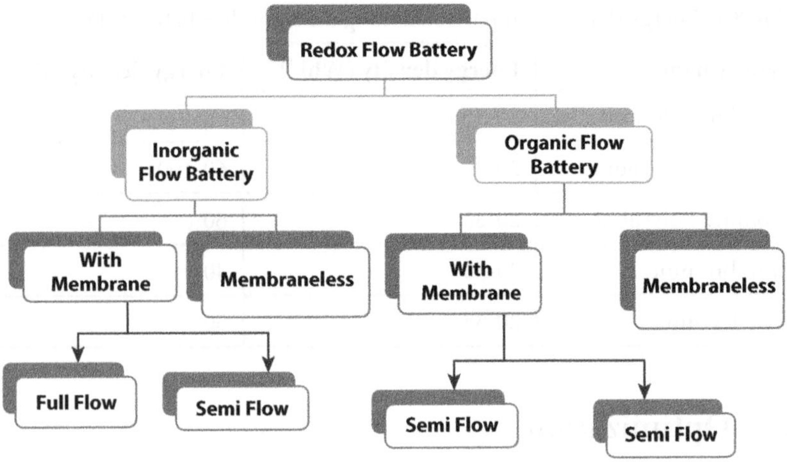

Figure 8.2 Classification of RFB.

8.5.1 Inorganic Redox Flow Battery

8.5.1.1 Vanadium Redox Flow Battery

In the early 1980s, Skyllas-Kazacos and coworkers proposed the vanadium–vanadium redox flow battery (VRB) at the University of New South Wales, Australia [19, 20]. The major inherent problem of redox flow batteries was the cross-contamination of different electrolytes that had been used in two half-cells and resulted in a gradual capacity loss, which was eliminated by the invention of VRB. Vanadium has been chosen because of its stability in various oxidation states and distinct half-cell potential which can produce cell storage in a practical sense. Here, in VRF five different oxidation states (V^{2+}/V^{3+} and V^{4+}/V^{5+}) of vanadium are in use (Figure 8.3). H_2SO_4 is used instead of HCl in VRF as a supporting electrolyte for both the half-cell redox couple. During overcharging H_2SO_4 produces oxygen in the cathode half cell which is environmentally accepted while the use of HCl produces chlorine gas in the positive half-cell, which is not accepted. The problem of the low solubility of V(V) compounds in acidic solutions, which limits the energy density, has been removed by using mild sulphuric acid as an electrolyte (< 5 and 2 M). In 2020 Cheng *et al.* deployed Prussian blue analog (PBA) $(VO)_6[Fe(CN)_6]_3$ as a capacity booster to rid of the earlier-mentioned issue [21].

Generally in RFB, all the active analysts remain in the electrolyte, but here $VOSO_4$ in mild H_2SO_4 is used to get active V^{4+} in a solution.

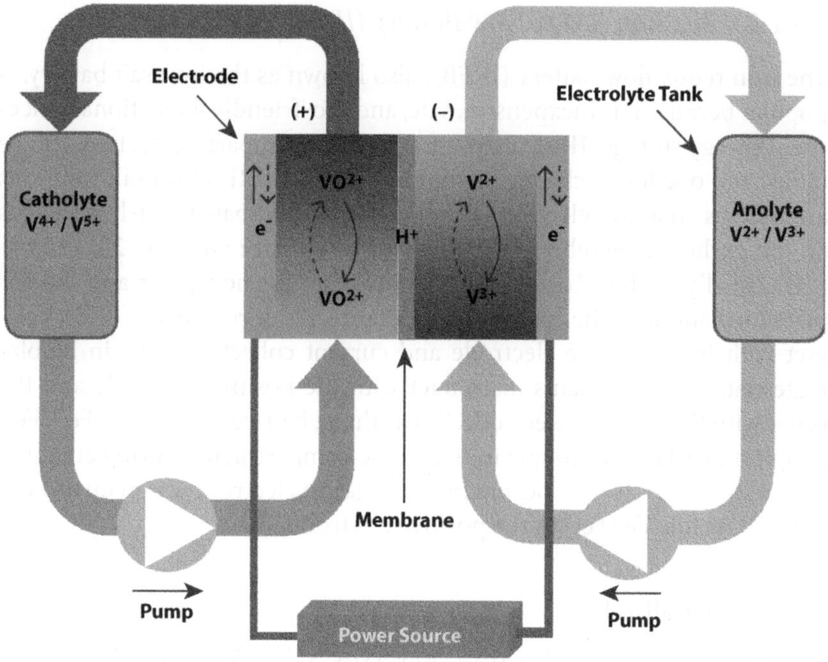

Figure 8.3 Schematic representation of the working protocol of vanadium redox flow battery.

As the system has only one active V^{4+} ion, it significantly reduces the cross-contamination of anolyte and catholyte [22–25].

Charging:
Cathode Half-cell:

$$VO^{2+} + H_2O - e = VO_2^{+} + 2H^{+} E^0 = +1.00 \text{ V versus SHE (Standard Hydrogen Electrode)}$$

Anode Half-cell:

$$V^{3+} + e = V^{2+} E^0 = -0.26 \text{ V versus SHE}$$

Overall cell reaction for charging:

$$VO^{2+} + H_2O + V^{3+} = VO_2^{+} + 2H^{+} + V^{2+}$$

$$E^0_{cell} = E^0_{cathode} - E^0_{anode} = (1.00 - (-0.26)) \text{ V} = 1.26 \text{ V}$$

8.5.1.2 The Iron Redox Flow Battery (IRFB)

The iron redox flow battery (IRFB), also known as the iron salt battery, is popular because it is inexpensive, safe, and eco-friendly for stationary electrical energy storage. IRFB also consists of two compartments, two storage tanks, and one ion exchange membrane. Here, in IRFB iron salt solutions are used as an active electrolyte, and carbon-based paper or felt is used as an ion exchange membrane for their higher surface area [26–28] (Figure 8.4). Usually, carbon-based materials are used for non-polar and bipolar plate formation. In the case of monopolar plates, it remains in direct contact with its respective electrode and current collector, while in bipolar plates either side remains in contact with the positive electrode and the other with the negative electrode. The cathode half-cell contains a Fe^{3+}/Fe^{2+} couple and a Fe^{2+}/Fe^{0} couple in the anode compartment. During charging oxidation of Fe^{2+} to Fe^{3+} occurs in the negative electrode and reduction of Fe^{2+} to metallic Fe (Fe^{0}) in the positive electrode.

Charging:
Cathode Half-cell:

$Fe^{2+} + 2e = Fe^{0}$ $E^{0} = 0.77$ V versus SHE (**S**tandard **H**ydrogen **E**lectrode)

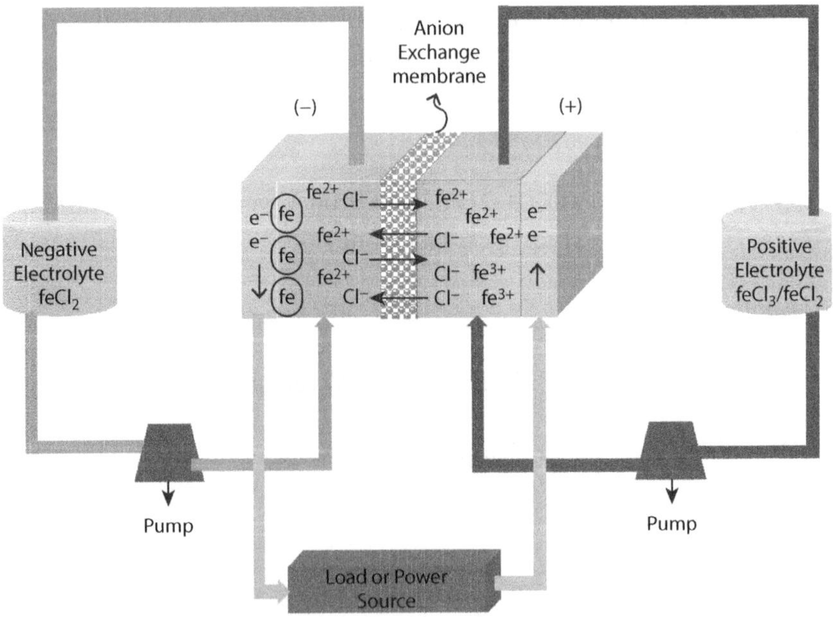

Figure 8.4 General formulation of IRFB.

Anode Half-cell:

$$2Fe^{2+} - 2e = 2Fe^{3+} E^0 = -0.44 \text{ V versus SHE}$$

Overall cell reaction for charging:

$$3Fe^{2+} = Fe^0 + 2Fe^{3+} \qquad E^0_{cell} = E^0_{cathode} - E^0_{anode} = (.77 - (-0.44)) \text{ V} = 1.21 \text{ V}$$

Low cost, higher stability, and availability of iron make IRFB a most promising alternative option for energy storage. However, it comes with obstacles like varying pH and H_2-production as a side reaction, which causes permanent efficiency loss of the battery. While charging, the reduction of Fe^{2+} to Fe^{3+} is accompanied by the liberation of an e^-, which is captured by the H^+ ion of an acidic solution and produces H_2 gas.

$$\text{Cathode reaction: } Fe^{2+} (aq) = Fe^{3+} + e$$

$$H^+ + e = H_2 (g)$$

The kinetics of the hydrogen evolution reaction depends on the concentration of H^+ or pH of the solution. Hence, at lower pH, H_2 gas production increases. In the due course of cell reaction, slowly pH increases at the negative electrode and this leads to the formation of iron hydroxide precipitation on the separator when the pH reaches above 3.5, resulting in a formation of resistance to the ionic transfer. To avoid insoluble iron hydroxide formation, the operational pH of IRFB needs to be lower than 3.5. On the other hand, at lower pH, the evolution of H_2 gas increases, which results in the permanent cessation of battery efficiency. It has been observed that at higher pH coulombic efficiency increases by manyfold. Various techniques have been applied to surpass the hindering reaction in both the cathode and anode compartment to smoothen the reaction kinetics. $FeCl_2$ has been chosen as a base electrolyte over $FeSO_4$ because chloride salt shows higher ionic conductivity and higher aqueous solubility (> 3M). On another note, an inclination was found towards the use of sulfate electrolytes because of their less corrosive and eco-friendly nature. Yu *et al.* [29] have used a highly concentrated $FeSO_4$/1-ethyl-3-methylimidazolium chloride electrolyte for IRFB. They used a minor amount of 1-ethyl-3-methylimidazolium chloride to enhance the aqueous solubility of $FeSO_4$ salt.

8.5.1.3 Polysulphide-Bromine Redox Flow Battery (PBBs)

Unlike vanadium flow batteries, here in PBBs different active species are stored in two compartments. Analytes contain an alkaline solution of sodium polysulphide, while an aqueous solution of NaBr a catholyte. During the charging process reduction from Br_2 to Br^- occurs in the cathode. In catholyte Br_2 remains as a Br_3^- while the combination of elemental sulfur and sulfide anions forms polysulfides in anolyte [30] (Figure 8.5). The ion-exchange membrane transports sodium ions between the positive and negative electrodes to balance and complete the battery circuit.

Charging:
Cathode compartment:

$$2NaBr = Br2 + 2Na+ + 2e2Na+ + (x\text{-}1)\ Na_2S_x + 2e = xNa_2S_{x\text{-}1},\ x = 2 - 4$$

Anode Compartment:

$$E^0_{Br2/\ 2Br-} = 1.087\ V \qquad E^0_{polysulfide/\ s2-} = -\ 0.428\ V$$

$$\text{Overall cell potential } E^0_{cell} = E^0_{cathode} - E^0_{anode} = 1.515\ V$$

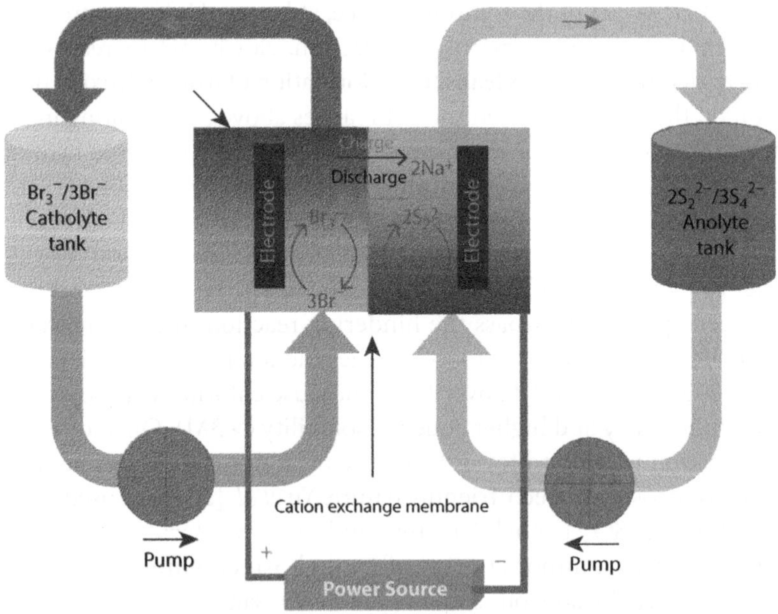

Figure 8.5 Working protocol of PBBs.

During the cell cycle, Br- ions release Br_2 and electrons in the cathode, and then Br^- and Br_2 combine to form a Br_3^- complex anion. Due to the concentration gradient and electric field, Br_3- can be transported from the cathode to the anode compartment resulting in self-discharge and leakage of energy efficiency. To reduce diffusion and enhance energy efficiency, Zhang *et al.* (2012) employed the polysulfide system in the cathode reaction [31].

As we all know, bromine is corrosive, thus corrosion-resistant electrodes need to be used as cathodes. In PBBS, carbon felt, graphite felt or carbon paper is used as a corrosion-resistant bromine electrode. Due to their high porosity carbon felt or carbon paper increases the activity of electrodes (Zhou *et al.* 2006) [32]. Shao *et al.* (2009) showed a large improvement in the electrochemical activity of the Br_2/Br- electrode by using carbon nanotubes as a cathode [33]. Activated carbon is another alternative material to use as an electrode at the anode half-cell. Calver *et al.* (1999) fabricated an electrode for an anode compartment by shuck-type activated carbon. Later in 2005, Zhao *et al.* showed a 20% enhancement in anode energy efficiency by replacing activated carbon with nickel or cobalt-coated carbon felt.

8.5.1.4 *Zinc-Polyiodide Redox Flow Battery*

There is continuous progress in the development of RFB but still, it has not reached the peak where the traditional battery can match the energy density or power with low-end lithium (Li) ion batteries. One of the high energy density RFBs is Zinc-polyiodide RFB, where the energy density is theoretically 322 Whl^{-1} at the solubility limit of ZnI_2 is 4500g.l^{-1}, 7.0 M (in aqueous solution). One of the prominent reasons behind the low energy density of RFB is the low aqueous solubility of electrolytes. Bin Li *et al.* demonstrated a Zn-polyiodide RFB, which has a discharge energy density of 166.7 Whl-1 with 5.0 M ZnI_2 electrolytes (Table 8.2, Figure 8.6) [34].

Table 8.2 The performance for other electrochemicals ZIB cell with the ZnI_2 (conc. of 3.5) M runs at different charge/discharge current density [34].

Current density ($mA\ cm^{-2}$)	OCV (V)	Charge voltage (V)	Average discharge voltage (V)	Average discharge energy density (WhL^{-1})
10	1.240	1.298	1.108	141.5
20	1.270	1.362	1.066	135
30	1.243	1.416	0.983	125.3

Hence, the efficiency of zinc-polyiodide RFB is more than seven times that of VRFB [35, 36]. Now coming to the construction of Zinc-polyiodide, is an example of an inorganic semi-flow hybrid battery system. Here, zinc

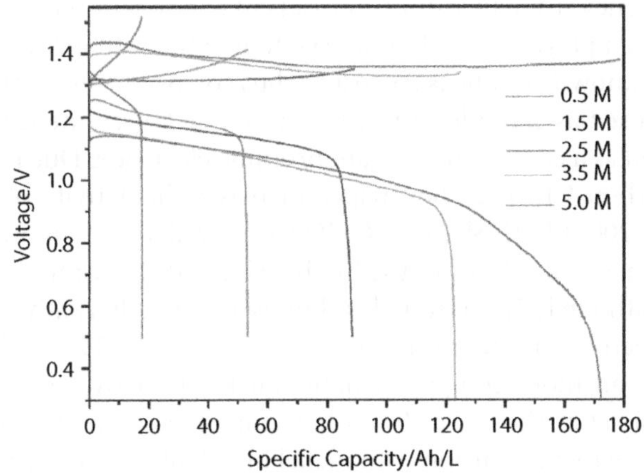

Figure 8.6 Voltagevs specific capacity graph at different ZnI_2 concentration where the current densities 20 mA cm^{-2} for 0.5-3.5 M and 20 mA cm^{-2} for 5M concentration of ZnI_2 [34].

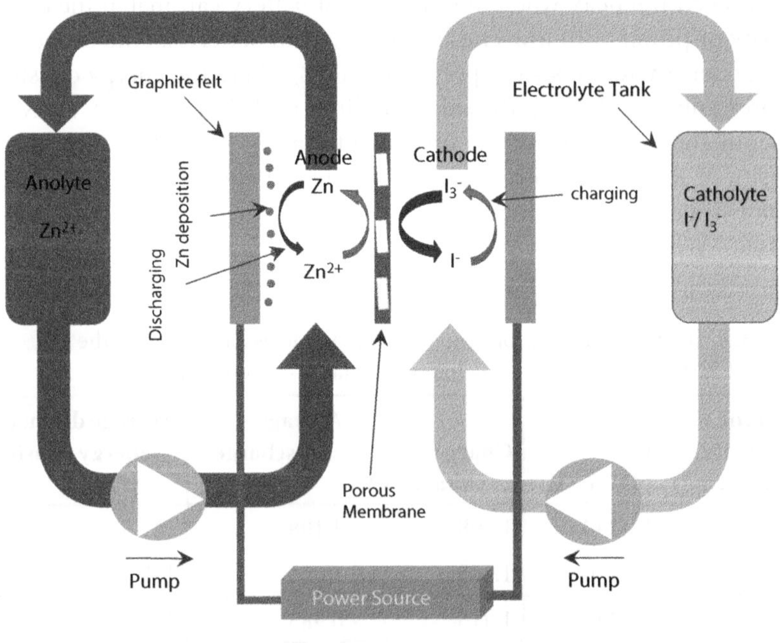

Figure 8.7 General working protocol of Zinc-polyiodide redox flow battery.

metal is deposited on anode compartments; that is why it is called semi-flow hybrid RFB. The anode and cathode compartments are Zn^{2+}/Zn and I_3^-/I^-, respectively (Figure 8.7). Triiodide has higher solubility in an aqueous solution than I_2 and hence triiodide remains in a predominant amount in an aqueous solution.

The Cell reaction of Discharging:
Cathode Half-cell:

$$I_3^- + 2e = 3I^- \quad E^0 = +0.536 \text{ V versus SHE (Standard Hydrogen Electrode)}$$

Anode Half-cell:

$$Zn = Zn^{2+} + 2e \quad E^0 = -0.76 \text{ V versus SHE}$$

Overall cell reaction for discharging:

$$I_3^- + Zn = 3I^- + Zn^{2+} \quad E^0_{cell} = 1.29 \text{ V}$$

Both cathode and anode contain ZnI_2 salt in an aqueous solution (Zn^{2+} and I^-). During the charging process Zn^{2+} ions accept 2 e-, form Zn metal, and get deposited at the anode, while on the positive side, polyiodide is formed. The reverse reaction occurs during the discharge process. Zinc ions have dual activity, one as an active species at the anode and two, as a charge transporter through the membrane. It has been observed that Zn-polyiodide RFB is one of the safest ones because there is no chance of inflammability, no risk from acidic solution, and the operating temperature is in the range of -20°C to 50°C (Table 8.3). The two major causes of efficiency reduction are i) transportation of electroplated Zn through the membrane and ii) Zn dendrite formation. The formation of Zn dendrite can be erased by adding ethanol (Table 8.3).

Table 8.3 The stability of catholyte with the presence of different additives (Ethanol (EtOH), ethylene glycol (EG)) at 100% SOC with different ZnI_2 concentration at different temperature in 10 days [34].

ZnI_2 (M)	Vol % of EtOH	50 °C	25 °C	0 °C	-10 °C	-20 °C
3.5	0	stable	stable	unstable	unstable	unstable
	25	stable	stable	stable	stable	stable
	25 (EG)	stable	stable	stable	stable	stable
2.5	0	stable	stable	unstable	unstable	unstable
	10	stable	stable	stable	stable	stable

8.6 Organic Redox Flow Battery

Inorganic compound-based RFB keeps on growing throughout the world but at the same time, considering the development cost of active redox materials, toxicity, cross-over property, and cost of the maintenance of ion exchange membranes provoke researchers to incline towards the use of the organic compounds in RFB [37, 38]. Organic compounds are very much known for their use in electrical devices like organic photovoltaic cells [39], transistors [40], metal-free batteries, etc. [41]. Figure 8.8 shows four types of organic compounds and their charge-storing mechanisms that have been applied in RFB. First, the pi-conjugated systems, for example, quinone, have been well explored as a redox-active material [42] and the second is based on phenazine [43], phenothiazine [44], or viologen [45–47], which are examples of fused ring aromatic molecules. They can remain either in cationic or anionic species. The third category is stable radical materials and the fourth one is organometallic complex materials [48–50]. Table 8.4 shows the comparison among different parameters of Aqueous Organic Redox Flow Batteries (AORFBs) [51].

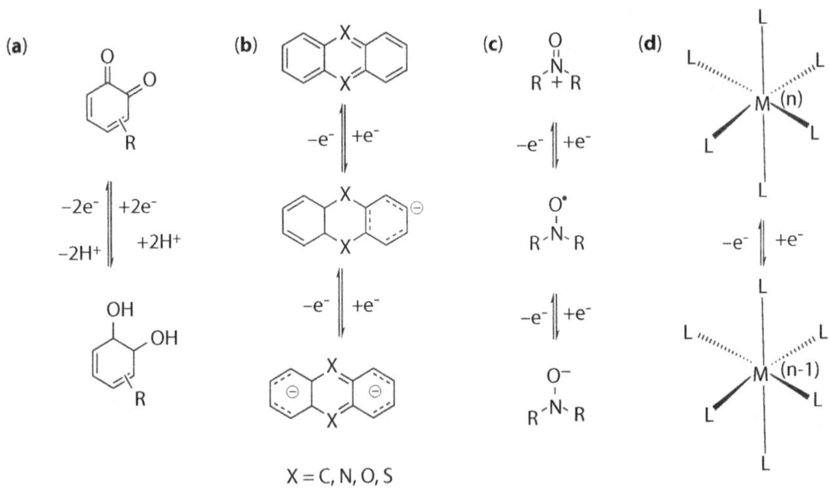

Figure 8.8 (a) Carbonyl based; (b) fused aromatic ring based; (c) stable radical based; and (d) organometallic based organic compound used in RFB.

Table 8.4 Comparison table (different parameters in AORFBs) [51].

Anolyte/catholyte (M)	E° (V)	Energy density[a] (WhL^{-1})	Achieved capacity[b] (%)	Peak power density (mW cm^{-2})	Capacity retention/ cycle (n° cycles)	EE (%) at current density- (mA cm^{-2})	Membrane	Cost (€ kW^{-1} h^{-1})[d]
2,7-AQDS/HBr/Br_2 1 M/3 M (acid pH)	0.87	25.7	69	0.6–1	99.84 (750)	76 (750)	N212	-
AQDS/$FeSO_4$ 0.33 M/0.67 M (acid pH)	0.62	5.5	70	0.134	99.999924 (500)[c]	70-75 (100)	N117	47
2,6-DBEAQ/$K_4[Fe(CN)_6]$ 0.5 M/0.3 M pH 12	1.05	6.5	85	0.24	99.9993 (250)	80 (100)	E620	44
2,6-DPPEAQ/$K_4[Fe(CN)_6]$ 0.5 M/0.4 M (pH 9)	1.0	7.7	97	0.16	99.99964 (480)	65 (100)	E620	44
DHPS/$K_4[Fe(CN)_6]$ 1.4 M/0.31 M (pH 14)	1.4	18.4	90	0.14	99.98 (500)	82 (100)	N212	–
ACA/$K_4[Fe(CN)_6]$ 0.5 M/0.4 M (pH 14)	1.2	4.6	86	0.35	99.9775 (400)	74 (100)	N212	–
BTMAP-Vi/TMAP-TEMPO 1.5 M/1.5 M (neutral pH)	1.19	22.1	Ca 90	0.099	99.985 (250)	52 (100)	AMV	–
BTMAP-Vi/BTMAP-Fc 1.3 M/1.3 M (neutral pH)	0.75	8.0	58	–	99.9989 (500)	–	DSV	–
BTMAP-Vi/BTMAP-Fc 0.75 M/1 M (neutral pH)	0.75	12.2	84	0.06	99.9943 (250)	66 (50)	DSV	–

(Continued)

Table 8.4 Comparison table (different parameters in AORFBs) [51]. (*Continued*)

Anolyte/catholyte (M)	E° (V)	Energy density[a] (WhL^{-1})	Achieved capacity[b] (%)	Peak power density ($mW\ cm^{-2}$)	Capacity retention/ cycle (n° cycles)	EE (%) at current density- ($mA\ cm^{-2}$)	Membrane	Cost (€ $kW^{-1}\ h^{-1}$)[d]
$(SPrN)_2V/NH_4[Fe(CN)_6]$ 0.9 M/0.9 M (neutral pH)	0.82	9.6	78	0.072	99.9997 (1000)	63 (40)	CSO	49
AQDS/$FeSO_4$ 0.33 M/0.67 M (acid pH)	0.62	5.5	70	0.134	99.999924 (500)[c]	70-75 (100)	N117	47
BTMAP-Vi/BTMAP-Fc 1.3 M/1.3 M (neutral pH)	0.75	8.0	58	–	99.9989 (500)	–	DSV	–
BTMAP-Vi/BTMAP-Fc 0.75 M/1 M (neutral pH)	0.75	12.2	84	0.06	99.9943 (250)	66 (50)	DSV	–
$(SPrN)_2V/NH_4[Fe(CN)_6]$ 0.9 M/0.9 M (neutral pH)	0.82	9.6	78	0.072	99.9997 (1000)	63 (40)	CSO	49

[a]Theoretical energy density calculated for 1:1 anolyte:catholyte ratio under given conditions.
[b]Achieved capacity was calculated as the percentage of capacity achieved from the theoretical capacity of limiting electrolyte.
[c]Long cycling at 200 $mA\ cm^{-2}$.
[d]Cost refers to electrolyte cost.

8.7 Membrane-Less RFB

The Laminar flow of liquid which is pumped from the storage tank to the reaction chamber is the prime requirement for a membrane-less battery. The laminar flow of two liquids keeps them separated and removes the importance of the membrane. In the construction of RFB, the membrane is the costliest one and very often gets corroded by frequent exposure to certain chemicals. For example, liquid bromine and H_2 form hydrobromic acid, which can destroy the membrane, hence, for such cases a membrane-less battery is demanded. In August 2013, one membraneless battery with quite good energy efficiency was reported, where the researchers used a small channel between two electrolytes.

8.8 Semi-Solid RFB

The name itself reveals that not all the charged particles are dissolved in the respective electrolytes. Here, in semi-solid redox flow batteries, there are suspended positive and negative particles in the carrier liquid. There is a separate tank for suspended particles, from which they are pumped through different pipes into the adjacent reaction vessel. Generally, a porous separator is constructed between the cathode and anode chamber (Figure 8.9).

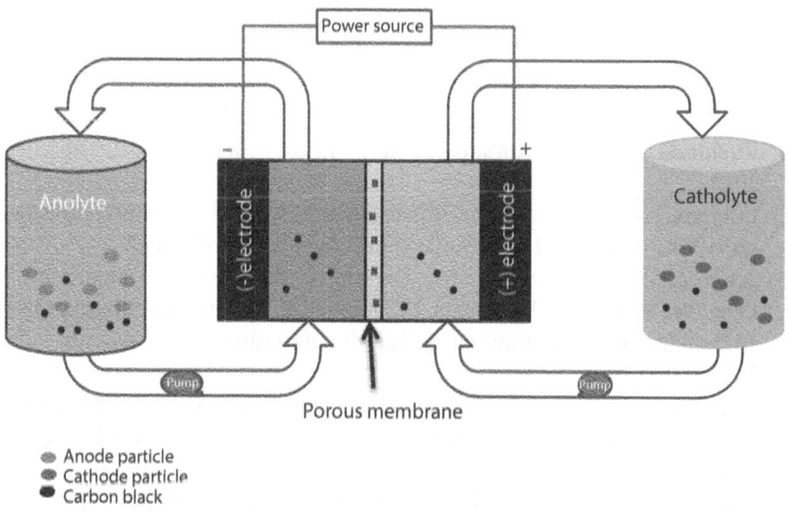

Figure 8.9 Schematic representation of semi-solid redox flow battery.

8.9 Conclusion

Globally, redox flow batteries (RFB) technology is the answer to the challenge of how to store energy. RFB is very safe to handle, has a high lifespan (around 20-25 years), almost 100% depth of discharge, and most importantly it is affordable. These make RFB an attractive choice for an energy storage system. Due to the large capacity of vanadium redox flow battery, it becomes a potential and promising energy storage grid to store the energy from wind and photovoltaic. One of the attractive features is that the long life span makes the iron redox flow battery more demanding than the Li-ion battery. Though several redox flow batteries are commercialized now, the large size of the electrolyte storage tank and high overall cost still inhibit the diffusion of redox flow batteries in the global market. Chemists are looking for sustainable organic polymeric materials to decrease the overall cost and tank size. R&D sectors are searching for an alternative strategy to minimize the tank size; hence it can be easily movable. There is a ray of hope that shortly, RFB will be a potential alternative for the energy storage system.

References

1. EU, The European Green Deal 640 final, Brussels, 2019.
2. Report on the Implementation of the Strategic Action Plan on Batteries: Building a Strategic Battery Value Chain in Europe, COM, 176 final, Brussels, 2019.
3. *The national academies summit on America's energy future: summary of a meeting.* N. R. C. Committee for the National Academies Summit on America's Energy Future, National Academies Press, Washington, DC, 2008.
4. Ritchie, H., Roser, M., Rosado, P., Energy, https://ourworldindata.org/energy, 2020.
5. Intergovernmental Panel on Climate Change. https://report.ipcc.ch/ar6wg3/pdf/IPCC_AR6_WGIII_SummaryForPolicymakers.pdf, 2022.
6. Scovronick, N., Budolfson, M., Dennig, F., Errickson, F., Fleurbaey, M., Peng, W., Socolow, R.H., Spears, D., Wagner, F., The impact of human health co-benefits on evaluations of global climate policy. *Nat. Commun.*, 10, 1, 2019.
7. Jacobson, M.Z., Delucchi, M.A., Guillaume, B., Zack A. F., B., Christa, C.H., Emma, F., Sean B., Morris, Diniana J. Y., P., Taylor A., V., Tim W., Y., 100% clean and renewable wind, water, and sunlight (WWS) all-sector energy roadmaps for the 50 United States. *Energy Environ. Sci.*, 8, 2093, 2015.

8. REN21, Renewables 2022 - Global Status Report (renewable energies), https://www.ren21.net/wp-content/uploads/2019/05/GSR2022_Full_Report.pdf, 2022.
9. REN21, Renewables 2021- Global Status Report (renewable energies) https://www.ren21.net/wp-content/uploads/2019/05/GSR2021_Full_Report.pdf, 2022.
10. Yang, Z., Zhang, J., Kintner-Meyer, M.C.W., Lu, X., Choi, D., Lemmon, J.P., Liu, J., Electrochemical energy storage for green grid. *Chem. Rev.*, 111, 3577, 2011.
11. Dunn, B., Kamath, H., Tarascon, J.-M., Electrical Energy Storage for the Grid: A Battery of Choices. *Science (80-.)*, 334, 928, 2011.
12. Wang, W., Luo, Q., Li, B., Wei, X., Li, L., Yang, Z., Recent progress in redox flow battery research and development. *Adv. Funct. Mater.*, 23, 970, 2013.
13. Soloveichik, G.L., Flow Batteries: Current Status and Trends. *Chem. Rev.*, 115, 11533, 2015.
14. Chakrabarti, M.H., Hajimolana, S.A., Mjalli, F.S., Saleem, M., Mustafa, I., Redox Flow Battery for Energy Storage. *Arab. J. Sci. Eng.*, 38, 723, 2013.
15. Skyllas-Kazacos, M., Rychick, M., Robins, R., All-Wanadium Redox Battery. United States Pat. US4786567A, 1988.
16. Weber, A.Z., Mench, M.M., Meyers, J.P., Ross, P.N., Gostick, J.T., Liu, Q., Redox flow batteries: A review. *J. Appl. Electrochem.*, 41, 1137, 2011.
17. Hall, D.M., Grenier, J., Duffy, T.S., Lvov, S.N., The Energy Storage Density of Redox Flow Battery Chemistries: A Thermodynamic Analysis. *J. Electrochem. Soc.*, 167, 110536, 2020.
18. Mohamed, M.R., Sharkh, S.M., Walsh, F.C., Redox flow batteries for hybrid electric vehicles: Progress and challenges. *5th IEEE Veh. Power Propuls. Conf. VPPC'09*, 551, 2009.
19. Rychcik, M., Skyllas-Kazacos, M., Characteristics of a new all-vanadium redox flow battery. *J. Power Sources*, 22, 59, 1988.
20. Colthorpe, A., Discovery and invention: How the vanadium flow battery story began. *Energy Storage News*, 2021.
21. Cheng, Y., Wang, X., Huang, S., Samarakoon, W., Xi, S., Ji, Y., Zhang, H., Zhang, F., Du, Y., Feng, Z., Adams, S., Wang, Q., Redox Targeting-Based Vanadium Redox-Flow Battery. *ACS Energy Lett.*, 4, 3028, 2019.
22. Cotton, F.A., Wilkinson, G., Murillo, C., Manfred, B., *Advanced Inorganic Chemistry* (6th ed.), Wiley-Interscience, 1999.
23. Atkins, P., Freeman, W.H., *Inorganic Chemistry* (5th ed.), 2010.
24. DOE/Pacific Northwest National Laboratory, Electric grid reliability: Increasing energy storage in vanadium redox batteries by 70 percent, *Science Daily*, 2011.
25. VRB Power Systems, https://vrbenergy.com/mw-class/2022.
26. Noack, J., Wernado, M., Roznyatovskaya, N., Ortner, J., Pinkwart, K., Studies on Fe/Fe Redox Flow Batteries with Recombination Cell. *J. Electrochem. Soc.*, 167, 160527, 2020.

27. Hruska, L.W., Savinell, R.F., Investigation of Factors Affecting Performance of the Iron-Redox Battery. *J. Electrochem. Soc.*, 128, 18, 1981.
28. Akilan, C., Nicol, M.J., Kinetics of the oxidation of iron(II) by oxygen and hydrogen peroxide in concentrated chloride solutions – A re-evaluation and comparison with the oxidation of copper(I). *Hydrometallurgy*, 166, 123, 2016.
29. Yu, S., Yue, X., Holoubek, J., Xing, X., Pan, E., Pascal, T., Liu, P., A low-cost sulfate-based all iron redox flow battery. *J. Power Sources*, 513, 230457, 2021.
30. Colin Tarrant, Special Feature: Electrical energy storage. *Power Eng. J.*, 122, 1999.
31. Zhang, S.S., Read, J.A., A new direction for the performance improvement of rechargeable lithium/sulfur batteries. *J. Power Sources*, 200, 77, 2012.
32. Zhou, H., Zhang, H., Zhao, P., Yi, B., A comparative study of carbon felt and activated carbon based electrodes for sodium polysulfide/bromine redox flow battery. *Electrochim. Acta.*, 51, 6304, 2006.
33. Shao, Y., Engelhard, M., Lin, Y., Electrochemical investigation of polyhalide ion oxidation-reduction on carbon nanotube electrodes for redox flow batteries. *Electrochem. Commun.*, 11, 2064, 2009.
34. Li, B., Nie, Z., Vijayakumar, M., Li, G., Liu, J., Sprenkle, V., Wang, W., Ambipolar zinc-polyiodide electrolyte for a high-energy density aqueous redox flow battery. *Nat. Commun.*, 6, 6303, 2015.
35. Martin, R.A.Y., Lecture Demonstrations Chemical Reactions II: Oxidation/Reduction. *J. Chem. Educ.*, 25, 495, 1948.
36. Svensson, P.H., Kloo, L., Synthesis, structure, and bonding in polyiodide and metal iodide-iodine systems. *Chem. Rev.*, 103, 1649, 2003.
37. Leung, P., Shah, A.A., Sanz, L., Flox, C., Morante, J.R., Xu, Q., Mohamed, M.R., Ponce de León, C., Walsh, F.C., Recent developments in organic redox flow batteries: A critical review. *J. Power Sources*, 360, 243, 2017.
38. Huskinson, B., Marshak, M.P., Suh, C., Er, S., Gerhardt, M.R., Galvin, C.J., Chen, X., Aspuru-Guzik, A., Gordon, R.G., Aziz, M.J., A metal-free organic-inorganic aqueous flow battery. *Nature*, 505, 195, 2014.
39. Barkhouse, D.A.R., Gunawan, O., Gokmen, T., Todorov, T.K., Mitzi, D.B., Yield predictions for photovoltaic power plants:empirical validation,recent advances and remaining uncertainties. *Prog. Photovoltaics Res. Appl.*, 20, 6, 2015.
40. Sirringhaus, H., 25th anniversary article: Organic field-effect transistors: The path beyond amorphous silicon. *Adv. Mater.*, 26, 1319, 2014.
41. Muench, S., Wild, A., Friebe, C., Häupler, B., Janoschka, T., Schubert, U.S., Polymer-Based Organic Batteries. *Chem. Rev.*, 116, 9438, 2016.
42. Ding, Y., Zhang, C., Zhang, L., Zhou, Y., Yu, G., Molecular engineering of organic electroactive materials for redox flow batteries. *Chem. Soc. Rev.*, 47, 69, 2018.

43. Hollas, A., Wei, X., Murugesan, V., Nie, Z., Li, B., Reed, D., Liu, J., Sprenkle, V., Wang, W., A biomimetic high-capacity phenazine-based anolyte for aqueous organic redox flow batteries. *Nat. Energy*, 3, 508, 2018.
44. Zhang, C., Niu, Z., Peng, S., Ding, Y., Zhang, L., Guo, X., Zhao, Y., Yu, G., Phenothiazine-Based Organic Catholyte for High-Capacity and Long-Life Aqueous Redox Flow Batteries. *Adv. Mater.*, 31, 1, 2019.
45. Janoschka, T., Martin, N., Hager, M.D., Schubert, U.S., An Aqueous Redox-Flow Battery with High Capacity and Power: The TEMPTMA/MV System. *Angew. Chemie - Int. Ed.*, 55, 14427, 2016.
46. Liu, T., Wei, X., Nie, Z., Sprenkle, V., Wang, W., A Total Organic Aqueous Redox Flow Battery Employing a Low Cost and Sustainable Methyl Viologen Anolyte and 4-HO-TEMPO Catholyte. *Adv. Energy Mater.*, 6, 1501449, 2016.
47. Orita, A., Verde, M.G., Sakai, M., Meng, Y.S., The impact of pH on side reactions for aqueous redox flow batteries based on nitroxyl radical compounds. *J. Power Sources*, 321, 126, 2016.
48. Lin, K., Chen, Q., Gerhardt, M.R., Tong, L., Kim, S.B., Eisenach, L., Valle, A.W., Hardee, D., Gordon, R.G., Aziz, M.J., Marshak, M.P., Alkaline quinone flow battery. *Science (80-.)*, 349, 1529, 2015.
49. Lin, K., Gómez-Bombarelli, R., Beh, E.S., Tong, L., Chen, Q., Valle, A., Aspuru-Guzik, A., Aziz, M.J., Gordon, R.G., A redox-flow battery with an alloxazine-based organic electrolyte. *Nat. Energy*, 1, 1, 2016.
50. Orita, A., Verde, M.G., Sakai, M., Meng, Y.S., A biomimetic redox flow battery based on flavin mononucleotide. *Nat. Commun.*, 7, 1, 2016.
51. Sánchez-Díez, E., Ventosa, E., Guarnieri, M., Trovò, A., Flox, C., Marcilla, R., Soavi, F., Mazur, P., Aranzabe, E., Ferret, R., Redox flow batteries: Status and perspective towards sustainable stationary energy storage. *J. Power Sources*, 481, 228804, 2021.

9

An Overview of Large-Scale Energy Storage Systems

Silas Saka

Department of Biotechnology, Institute of Technology, GITAM (Deemed to be University), Visakhapatnam, Andhra Pradesh, India

Abstract

The constant logarithmic progression of finance and technology has increased energy usage. Due to global warming and temperature fluctuations, the requirement for temperature regulators has increased, which in turn increased the energy demand. Energy storage is the method of converting an energy form to a storable type which could then be used as electrical energy at the time of need. Many countries have started the use of stored energy for various applications. Accordingly, the study of energy storage systems is indispensable to implement and improve the utility of stored energy. Energy can be stored for future use in many forms like mechanical and thermal, chemical and electrical. Earlier energy storage systems were used only for power supply. Nowadays, energy storage systems can also be utilized for various other applications such as load levelling, peak shaving, and many more. Energy storage systems are creating new commercialization by linking consumers and producers. The large-scale usage of energy sources is increasing day by day. A proper understanding of these energy storage systems is essential for their proper utilization. Hence, this chapter deals with every possible aspect of various energy storage systems needed for their widespread application.

***Keywords*:** Pumped hydro storage, compressed air energy storage, flywheel energy storage, electrostatic energy storage, capacitors, batteries, fuel cells, power plants

9.1 Introduction

The constant logarithmic progression of finance and technology has increased energy usage. The International Energy Agency (IEA) reported

Email: ssaka@gitam.edu

Inamuddin and Tariq Altalhi (eds.) Advanced Redox Flow Technology, (199–244) © 2024 Scrivener Publishing LLC

that cosmopolitan energy usage rose in 2018 from 900 TWh to 26,700 TWh [1]. Due to global warming and temperature fluctuations, the requirement for temperature regulators has increased, which in turn increased the energy demand. Till 2018, the majority of energy requirements were met by burning fossils (approximately 64%). The process has led to an increase in global temperature with the release of enhanced CO_2 [2]. Although the use of fossils is the major source of energy, the importance of clean energy has been recognized. To meet the vision of "Paris Agreement - carbon neutral" by 2060, the use of inexhaustible energy is receiving prominent consideration [3, 4]. According to 2018 data, renewable energy usage has been scaled by 7% and measures one-quarter of the whole cosmopolitan energy requirement [1, 2]. The mean yearly variation in the electrical energy supply of 2018 is highlighted in Figure 9.1.

The use of renewable energy improves economic status alongside environmental advantages [5]. The sporadic nature of these types of energy affects their utility in terms of stability, dependability, and durability. The work to find the path to solve this problem has been conducted. The possible solution focused on management, energy storage system, connecting networks, and many more. It was emphasized that energy storage could be the best potential solution to the above-mentioned problems [6].

Energy can be stored for future use in many forms like mechanical and thermal, chemical and electrical. The dependability and durability of these renewable energy sources can be improved by depicting energy production profile and managing accordingly [7]. The excess energy can also be conserved to avoid energy leakage. The recent use of stored energy is an electric vehicle used for transportation. Many countries have started the use of stored energy for various applications. Accordingly, the study of the energy storage system is indispensable to implement and improve the utility

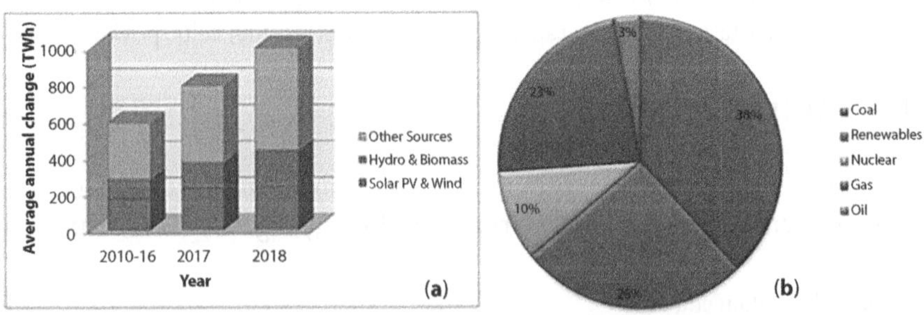

Figure 9.1 (a) The average annual change in electrical energy generation and (b) The electricity generation mix in 2018 [1].

of stored energy. The upcoming sections will discuss progression, categorization, implementation, commercial prototype, and environmental repercussions in detail. The chapter also summarizes the guidelines, blockades, prominent solutions, and also prospects of the stored energy system.

9.2 Progression of Energy Storage Method

Energy storage systems became active during the 19th century and travelled a way to gain their shape. Earlier in this period, humans were using wood as a source of energy for various works but not storing energy for use in times of need. The first energy storage device made was batteries. The term was first used by Benjamin Franklin in 1749, and batteries were first made by Alessandro Volta in 1800. Volta's battery was made of copper and zinc filled with electrolytes [8, 9]. This created a revolution in the field of energy storage which ultimately led to the development of various other battery types. The 20th century was known for the development of various battery types. The development of various storage systems is presented in a sequential array in Table 9.1.

Table 9.1 The evolution of energy storage systems (ESSs) in chronological order

Year	ESS	Introduction	Reference
Oldest	Bio fuels	Biofuels are in use since ancient times, in the form of wood and other types of solid biomass.	[10]
1839	Fuel cells	William Grove, a chemist, physicist, and lawyer invented the fuel cell in 1839.	[11]
1859	Lead acid battery	Invented in 1859 by French physicist Gaston Planté, lead acid battery is the earliest type of rechargeable battery.	[12]

(Continued)

Table 9.1 The evolution of energy storage systems (ESSs) in chronological order. (*Continued*)

Year	ESS	Introduction	Reference
1882	PHS	In 1882, the world's first hydroelectric power plant began operating in the United States along the Fox River in Appleton, Wisconsin.	[13]
1883	FES	The first FES used purely for ES was made of steel. It was developed by John A. Howell in 1883 for military applications.	[14]
1884	Flow batteries	French engineer Charles Renard invented flow batteries in 1884 to power an airship with an electric propeller.	[15]
1899	Nickel Cadmium battery	Nickel cadmium battery was invented by Waldemar Jungner in 1899. Further improvements occurred in 1947 which led to the modern sealed NiCd battery.	[16]
1899	Nickel Zinc battery	Nickel zinc battery was first invented in 1899. Its commercial production started in 1920.	[17]
1901	Nickel iron battery	The nickel iron battery was developed by Thomas Edison in 1901, and used for electric vehicles, such as the Detroit Electric and Baker Electric	[18]
1960	Sodium sulphur battery	Sodium Sulphur battery was first discovered by Ford Motor Company in 1960 to power early-model electric cars.	[19]
1967	Nickel metal hydride battery	Research on nickel-metal-hydride battery started in 1967.	[16]

(*Continued*)

Table 9.1 The evolution of energy storage systems (ESSs) in chronological order. (*Continued*)

Year	ESS	Introduction	Reference
1968	Sodium metal halide battery	Also known as ZEBRA batteries, sodium metal halide batteries were first introduced in South Africa in 1968.	[20]
1969	Super magnetic	SMES was first proposed as a method of diurnal storage in the year 1969.	[21]
1970	Nickel hydrogen battery	The Nickel-hydrogen battery was developed Comsat laboratories and Tyco laboratories, sponsored by Intelsat.	[22]
1978	Super capacitor	Research on supercapacitors began in the 1957 by General Electric Company. Followed by many episodes of development, the supercapacitor was marketed in 1978.	[23]
1978	CAES	The first utility-scale CAES project was the 290 MW Huntorf plant in Germany using a salt dome, which started in 1978.	[24]
1980	Li-ion battery	In 1980, American physicist Professor John Goodenough invented Li-ion battery.	[29]
1980	Li-polymer battery	The discovery of the lithium polymer battery came in the 1980s. The first commercial Li-ion cell was of Sony in 1991.	[25]
2021	Solar fuels	Driven by environmental concerns, solar fuels are recently gaining attention, and are still under research and development.	-

9.3 Categorization of Energy Storage System

Energy demand has resulted in the development of various types of batteries to increase their durability, utility, and dependability. These storage systems were categorized depending on their function, reaction period, storage time, and many more factors [26]. Figure 9.2 summarises the different storage systems and their types which are further elaborated on in the following sections.

Energy can be preserved for future use in many ways, mechanical, electrical, chemical, and thermal. It can also be stored by combining any two forms.

9.3.1 Mechanical Energy Storage

The classical form of power could be conserved effortlessly for years. Mechanical energy is easy to convert into other energy forms [3]. It is of two types: potential and kinetic energy. Potential energy is of two types:

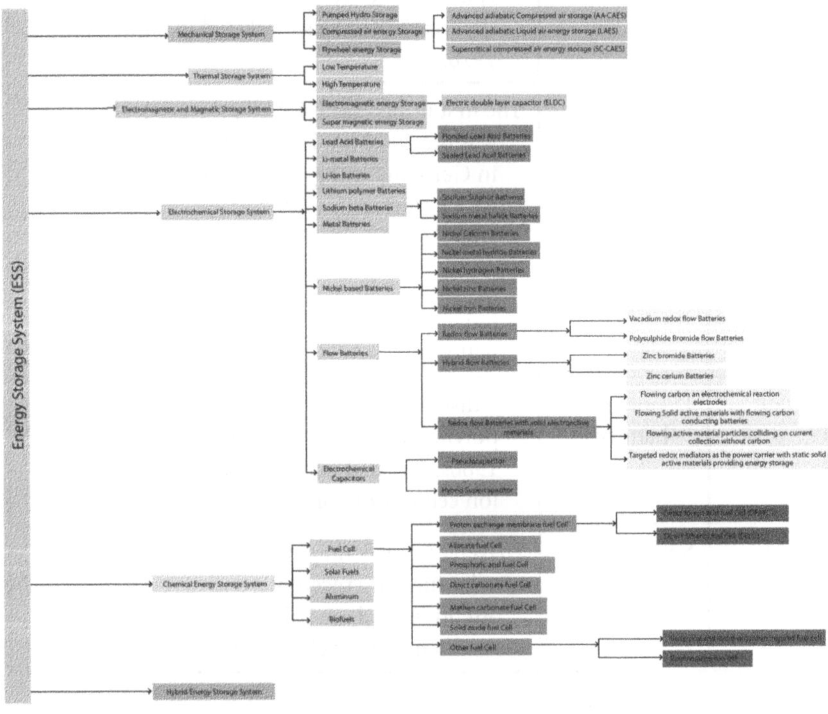

Figure 9.2 Classification of ESSs as described.

pressurized gas and forced spring. Kinetic energy is the energy that is stored in moving bodies. Compressed air storage and pumped storage hydropower are used for potential energy. For the storage of kinetic energy, a flywheel storage system is used.

Pumped hydro-storage was first used in Switzerland (1944). It has a large storage capacity of 3000 MW [27]. It can store 96% of total electric energy and has a 3% energy production potential [28]. Its efficiency is 77-85%. It is highly efficient, cheap, and durable. These are used in weight-carrying and shredding [27]. It is a level-up of the traditional hydroelectric power plant. The increased need for an electricity storage system using pumped hydro storage has been shown in Figure 9.3. The difference is that in the traditional system, water is contained in a reservoir and pumped to the turbines for its conversion to electricity, whereas in pumped hydro-storage the water can be multiple times. It uses two reservoirs, one above the other. From the upper tank water is released to the turbine and returned to the lower tank and utilized when the demand is up. A schematic representation of the pumped hydro storage system has been demonstrated in Figure 9.4.

From a given quantity of water, the energy generated is proportional to the head, which is the space between the upper reservoir and the turbines. Hence the energy generated will be directly proportional to the known quantity of water for a larger head. The amount of energy necessary to propel the aqua from the container at the bottom to the reservoir at the top is given by

$$P = \gamma Q \Delta H$$

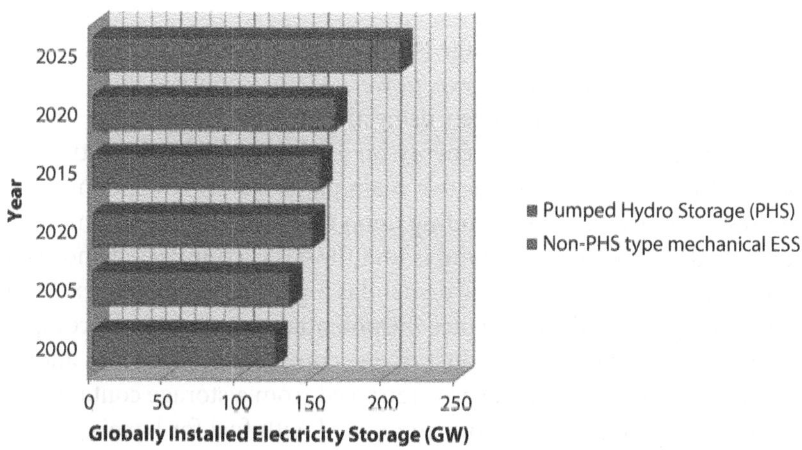

Figure 9.3 Cosmopolitan hike in pumped hydro storage system from 2000-2025 [3].

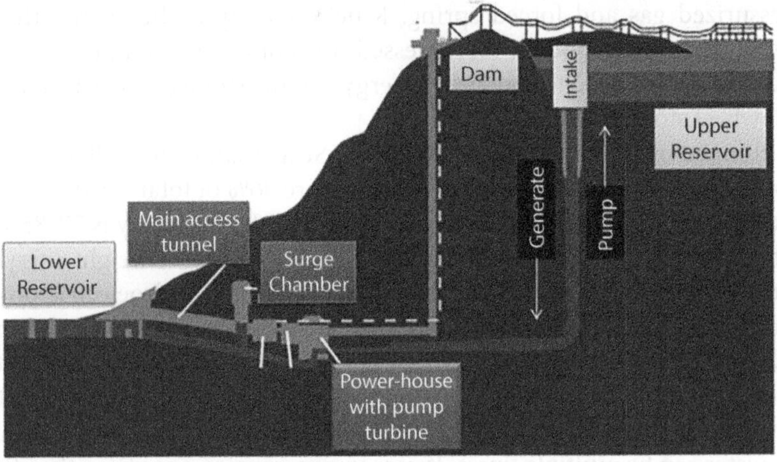

Figure 9.4 A simple demonstration of pumped hydro storage system [27].

Where,

γ; the definite mass of the fluid (lb/ft^3)

Q; Flow rate [ft^3/s)

ΔH; Difference in the head (ft)

The power generated from the turbine when the efficiency is η is shown as,

$$P_{generating} = \eta\gamma Q\Delta H$$

The currently operated pumped hydro storage projects of the US are listed in Table 9.2.

Compressed air energy storage is a mechanization which applies dense and pressurized gas to conserve power. Many countries like the UK, Germany, and France have been using the storage system since the 19th century [27]. It can also store energy on a large scale. Its capacity ranges from 35-300 MW. It constitutes two main parts: a storage container and a compressor or expander. Depending upon the need, the gas is compressed and expanded. When the demand is low, the gas is compressed and stored in a storage container. When the demand is high, the gas is again expanded and the energy is used. To enhance the potential, sometimes two compressors are applied where the product of one compressor is put into another. During need the compressed gas is retrieved from a storage container and inserted into a combustion room and mixed with fuel for burning. Figure 9.5 is a flowchart representation of a compressed air energy storage system.

Table 9.2 List of currently operated hydro storage projects of US [30].

Project	Initial operation	Installed capacity (MW)	Hours of storage	Energy storage (MWh)	Average gross head (feet)	Water conduit length (feet)	Length to head ratio; L/H	Number of existing reservoirs/ lakes
Muddy Run	1967	855	14.3	12,200	386	1290	3.3	1
Yards Creek	1965	330	8.7	2894	723	3700	5.1	0
Tam Sauk	1963	350	7.7	2700	809	7003	8.7	0
Northfield	1972	1000	10.1	10,100	772	6790	8.8	1
Cabin Creek	1967	280	5.8	1635	1159	4340	3.7	0
Seneca	1969	380	11.2	3920	736	2520	3.4	1
Ludington	1973	1888	9.0	15,000	337	1252	3.7	1
Blenheim Gilboa	1973	1030	11.6	12,000	1099	4355	4.0	0
Bath Country	1985	2100	11.3	23,700	1180	9446	8.0	0
Fairfield	1978	512	8.1	4096	163	2120	13.0	0
Bear Swamp	1974	540	5.6	3019	725	2000	2.8	0
Jocassee	1973	628	93.5	58,757	310	1700	5.5	1
Raccoon Mountain	1978	1370	24.0	33,000	968	3650	3.8	1
Helms	1984	1200	118.0	14,200	1645	20,519	12.5	2

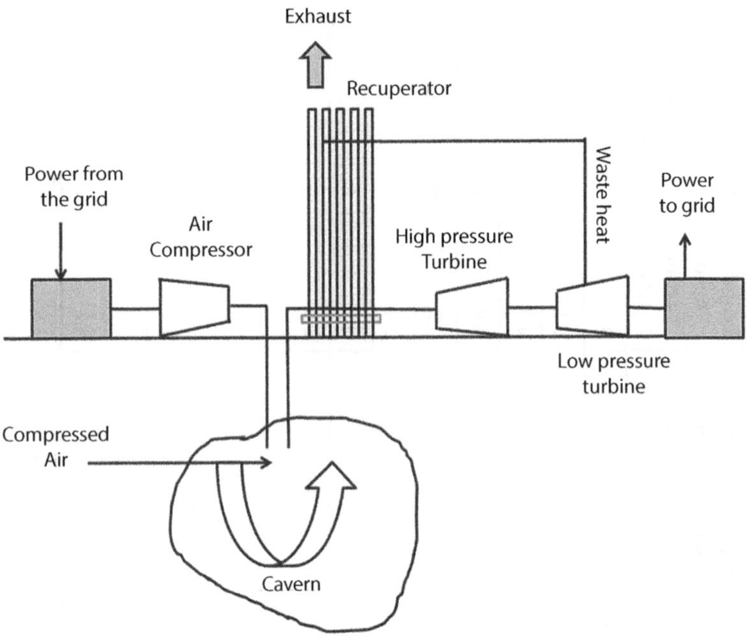

Figure 9.5 A diagramatic representation of compressed air energy storage (CAES) plant.

Imagine Q to be the heat flow, work, heat & energy fluxes for every element of the system are determined using the 1st and 2nd law of thermodynamics. The roundtrip emissions intensity (GI_{plant}), energy efficiency ($\eta_{storage}$), energy density, heat rate (HR) and work ratio (WR), can be determined. with the following equations:

$$\eta_{storage} = \frac{W_{TB}}{-W_{CM} + X_{NG} - W_{electrolysis}}$$

$$WR = \frac{-W_{CM} - W_{electrolysis}}{W_{TB}}$$

$$HR = \left(\frac{m_{NG} LHV_{ng}}{W_{TB}}\right)\left(\frac{3.6GJ}{MWH}\right)$$

$$GI_{plant} = (HR)(GI_{NG})$$

$$\rho = \frac{W_{TB}}{V}$$

Advanced Adiabatic Compressed Air Energy Storage is environmentally friendly. The heat generation and excess CO_2 liberation into the surrounding by the compressed air energy storage mechanism is overcome by this methodology. It is not dependent on the combustion process as heat storage is amalgamated with the existing Compressed Air Energy Storage system. Its efficiency is 70% and worked well with small and medium-level applications. Figure 9.6 is a flow chart representation of the advanced adiabatic compressed air energy storage (AA-CAES) system. The released heat is conserved in the adiabatic tank and is utilized for the expansion process. The adiabatic process does not exchange heat with the surrounding. It releases energy in the semblance of work. Air first blows into air storage space and then it enters into a heat chamber for heating.

The storage system that earned attention is liquid air energy storage. It uses liquid instead of air for compression. The system is profitable concerning space demand. It utilizes "cryogenic liquefaction technology." It is dual-staged. In phase one, the air is liquefied under high pressure and preserved in the cryogenic container. In the second phase, liquid air is expanded to run motors. Its storage capacity is 73%.

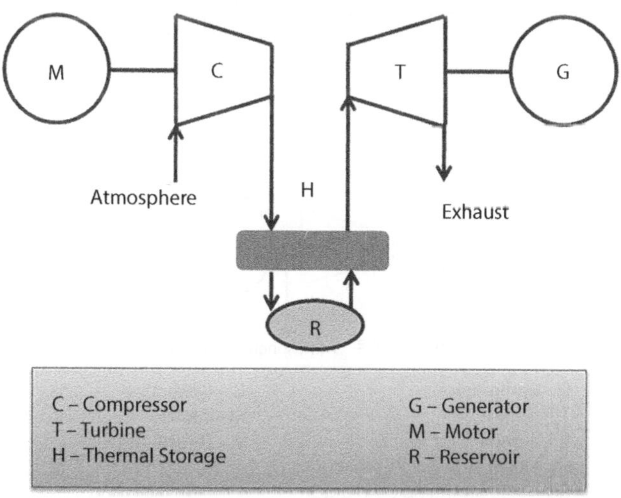

Figure 9.6 Flowchart representation of advanced adiabatic compressed air energy storage (AA-CAES) system [31].

A supercritical compressed air energy storage system is highly advanced and environmentally friendly. It is highly efficient. It is the fusion of liquid air energy storage and Advanced Adiabatic Compressed Air Energy Storage. In this system, the air is compressed to a supercritical state with high pressure (>37.9 bar) and temperature (> 132 K). This compressed air is stored as liquid air while moving through heat exchange. The heat exchanger gathers the heat from the air and uses it to produce energy [37]. The air can be liquefied by either of the two processes: throttle liquefaction valve (it utilizes the cooling outcome of air to generate liquefied air) or liquefied expander. An illustration of the SC-CAES system is shown in Figure 9.7, and the various types of CAES systems working currently are compared in Table 9.3.

Flywheel energy storage was first used by John A. Howell in 1883 for military application [14]. It is highly efficient with a high power range. Its energy storage capacity is 500 MJ. It can be used in the military, satellites, transports and many more [33]. It has a longer cycle and high response time. It is useful for large-scale power supply. It is based on the kinetic energy of a moving barrel to preserve mechanical energy. The internal and external flywheel energy storage system has been highlighted in Figures 9.8 (a) and (b).

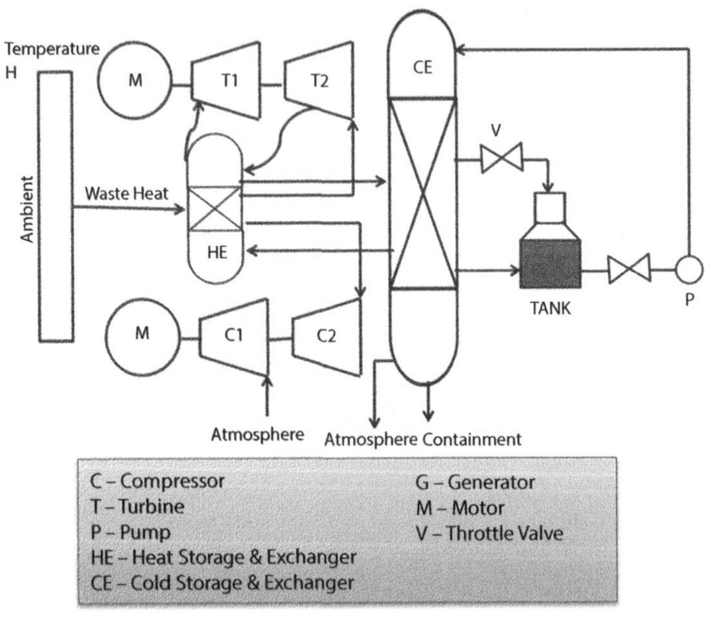

Figure 9.7 An illustration of SC-CAES system [31].

Table 9.3 Difference between various CAES systems [31].

Technology	Energy density (Wh/L)	Power rating (MW)	Storage duration	Lifetime (Years)	Discharge time	Cycling times (Cycles)
Large CAES	2-6	110-290	Hours - Months	20-40	1-24 + Hour	8000-12000
Small CAEs	2-6	0.003 & 3	Hours - months	23+	Upto - Hour	Test 30,000
AA-CAES	2-6	110 & 290	Hours - Months	20-40	1-24 + Hour	-
SC-CAES	8-24	110 & 290	Hours - months	20-40	1-24 + Hour	-
LAES	8-24	0.3 & 2.5	-	20-40	1-12 + Hour	-

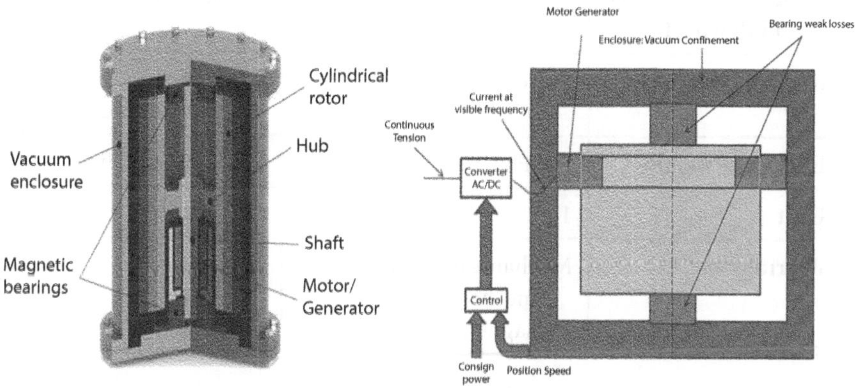

Figure 9.8 (a) External view of flywheel energy storage system; (b) the internal mechanism on which the internal flywheel mechanism is based on [32].

To protect from frictional loss the system is shafted on low pressure or vacuum. The system rotates at high speed to generate kinetic energy. It is then used to work the motor during discharging state [34]. Based on rotational speed, it is classified in two ways: lower (6000 rpm) and higher speed (10000-1000000 rpm) flywheel energy storage. The difference between low- and high-speed flywheel storage energy systems has been presented in Table 9.4. High-speed can be used for aerospace applications.

Rotor geometries affect the energy density as well as specific energy which is demonstrated with the use of K (shape factor) [33]. It predicts the rotor's capacity to preserve energy [35].

Table 9.4 Difference between -low-speed and high-speed flywheel energy storage system (FESS) [34].

Characteristic	Low-speed FESS	High speed FESS
Flywheel Material	Steel	Composite Materials: glass and C fibers
Confinement Atmosphere	Partial Vacuum or Light gas	Absolute Vacuum
Electrical machine	Asynchronous, permanent magnet synchronous and reluctance machines	Permanent magnet synchronous and reluctance machines
Integration of electrical machine and flywheel	No or partial integration	Full or partial integration
Main applications	Power quality	Traction and aerospace industry
Enclosure weight	2 X Flywheel weight	1/2 X Flywheel weight
Cost	1	5
Bearings	Mechanical or mixed (mechanical and magnetic)	Magnetic

The maximum specific energy can be demonstrated by the equation

$$\frac{E}{m} = K\frac{\sigma_{max}}{\rho}[J/kg]$$

and the highest energy density is

$$\frac{E}{V} = K\sigma_{max}[J/m^3]$$

The difference in K values for different flywheel geometries was represented in Figure 9.9.

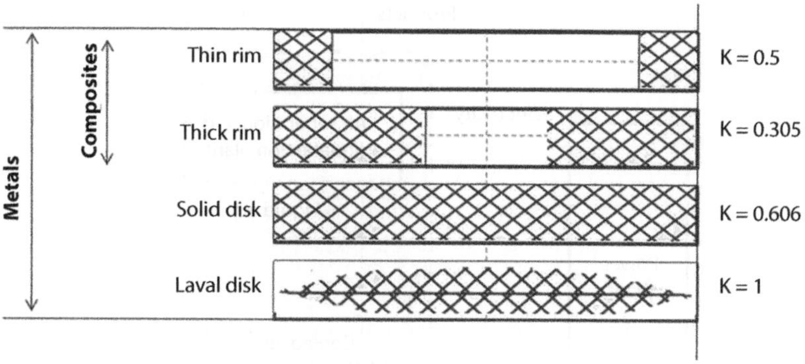

Figure 9.9 Different K values for different flywheel cross sections [34].

9.3.2 Thermal Energy Storage

Heat can be conserved in various ways, such as reversible thermochemical reactions, sensible heat and latent heat. This system can be applied for temporary, lasting storage and reuse [32]. The heat is stored by raising or lowering the temperature of a medium which can then be utilized to generate power [36]. It is reliable, economic, and efficient. It is used in industries and factories. The system is eco-friendly. It is used in collaboration with the solar photovoltaic system. This way large quantities of solar energy can be efficiently preserved and used when needed. The system can be divided into two categories: low-temperature thermal energy system and high-temperature thermal energy system.

In the low-temperature type, the heat is stored within a material and can be utilized from it. For cooling, chilled water or ice can be used. One of its special types is a cryogenic energy storage system, which uses liquid nitrogen (cryogen) for cooling and is generated during the off-peak time and reproduced during the high-peak time. The sub-freeze material is formed with the help of renewable energy or mechanical energy or a moving turbine. During need, the cryogen is heated to generate electricity. It has a high energy of one hundred to two hundred (approx) Wh/kg. It is durable and eco-friendly. Figure 9.10 presents the diagrammatic representation of cryogenic energy storage (CES).

In the high-temperature type, the energy is conserved in the form of sensible or latent heat. Sensible heat is the form of energy in which a body absorbs or releases heat by changing its temperature not affecting its volume, pressure or pressure. Latent heat is the form of energy in which an object consumes or emancipates heat by varying its state, not temperature.

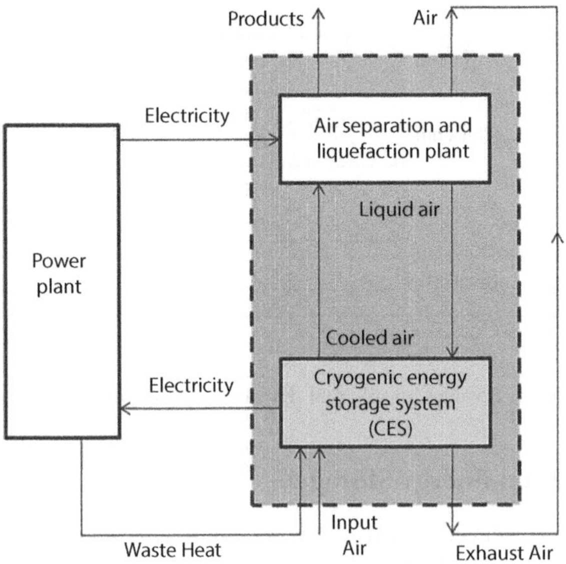

Figure 9.10 Diagramatic representaion of cryogenic energy storage (CES) [32].

The capacity of anybody to store or release heat energy is based on its heat capacity and specific heat [3]. The heat can be conserved in two forms, sensible and latent heat; the difference between the two has been graphically represented in Figure 9.11.

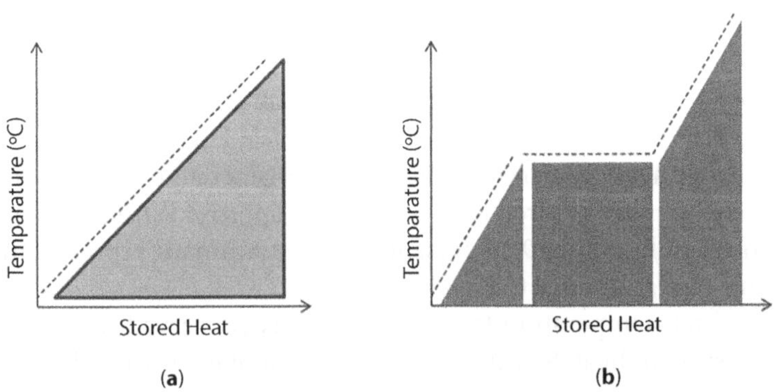

Figure 9.11 Graphical representation of storage forms for thermal energy (a) Energy storage in the from of sensible heat, (b) Energy storage in the form of latent heat [32].

Factors such as the change of temperature, specific heat and mass of storage material that influence the amount of heat stored are represented by

$$Q_s = \int_{t_i}^{t_f} \int mc_p dt = mc_p(t_f - t_i)$$

where,

Q_s; the amount of heat storage, (J); t_f; the final temperature, in (°C); t_i; the initial temperature, (°C); C_p; the specific heat, (J/(kg. K)); m ; the mass of heat storage medium, (kg).

9.3.3 Electrostatic and Magnetic Energy Storage System

With the help of supercapacitors and superconducting magnets, energy can be conserved in these forms. These are useful in generating high power and huge supply [3]. Electrostatic energy storage utilizes capacitors to conserve electrical energy among two conductive metallic plates gapped by an insulator material [37]. The gap between two plates determines the capacity of any capacitor. It also relies on the dielectric constant of the insulating substance as well as on the dimension of the transmitting plates. The quantity of stored energy relies directly on the capacitor. The capacitors can be charged instantaneously. It poses a long life cycle and low energy density. The system is useful for small-scale applications [37]. The advanced form of capacitors is supercapacitors. It has high charge density and storage potential. it also has a long life of 12 years. It can have very high output when connected with lead-acid batteries. In superconductors activated carbon in the form of electrolytes is used to separate conductive plates. Due to ion conduction, supercapacitors have a high specific area and energy density [37]. Supercapacitors use carbon nanotubes which further enhance their specific area and increase their energy density. Figure 9.12 highlights the diagrammatic representation of a conventional capacitor storage system.

The amount of energy conserved in an electric field of a supercapacitor where V is the voltage and C is capacitance across the capacitor is given by

$$W_{sc} = \frac{CV^2}{2}$$

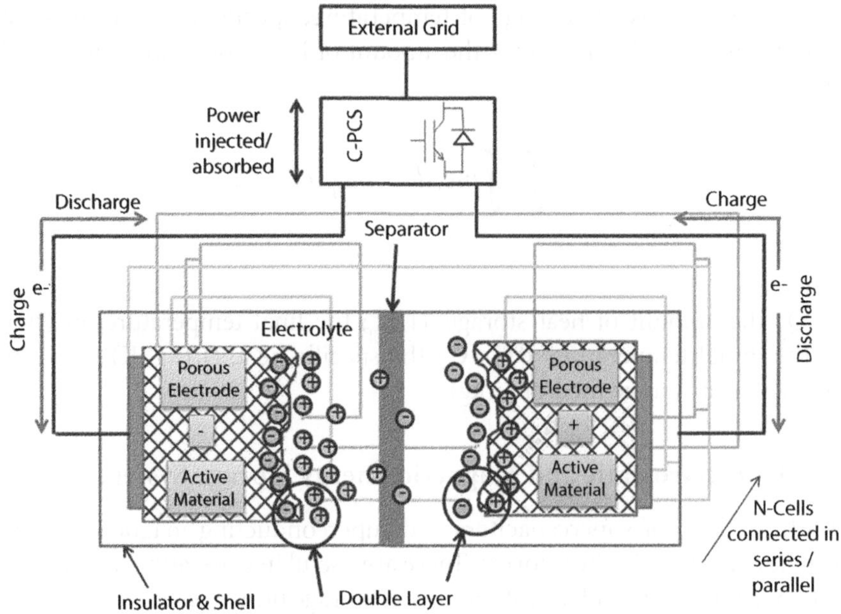

Figure 9.12 Diagrammatic representation of conventional capacitor storage system [32].

The above equation illustrates the accumulated total energy in a capacitor. But, in realistic usage, half of the power fall from the insignificant value is adequate during off-peak. Hence, the conserved energy in a capacitor will be described by the equation.

$$W_{sc} = \frac{c}{2}\left[V^2 - \left(\frac{v}{2}\right)^2\right] = \frac{3}{8}CV^2$$

Electric double-layered capacitors have more surface area of electrodes, which increases their charge density. In this system, electrodes are drowned in aqueous electrolytes and separated with the help of an ion-permeable separator. It helps to avoid short circuits. The energy is preserved in the form of ionic energy as ions are captivated from the electrolyte by the electrode [38], the same has been represented in Figure 9.13. Various electrodes, electrolytes and capacitance materials are used for storing energy in an electric double-layer capacitor. Table 9.5 presents the difference among various electrolytes and electrodes used along with their capacitance.

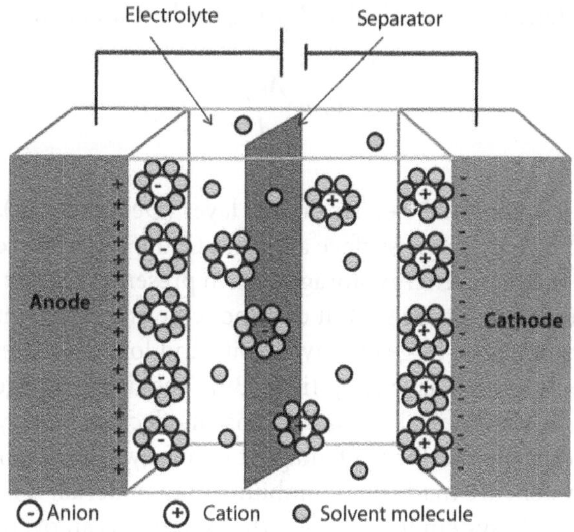

Figure 9.13 An illustration of electric double layer capacitor (EDLC) [37].

Table 9.5 Electrode, electrolytes, and their specific capacitance material used for different electric double layer capacity [37].

Electrode material	Electrolyte	Specific capacitance
AC Fibers	KOH	180-210
Mesoporous Carbon	KOH	180
Graphite and Reduced Graphene oxide (rGO)	Tetraethylammonium Tetrafluoroborate	10-150
Carbon/Aerogels xerogels	Aqueous (NaOH/KOH)	40-200
Carbon Black	Aqueous (NaOH/KOH)	<300
Carbide-Driven Carbon	Ionic Liquids (KCl/NaCl)	100-150
CNT	Aqueous (NaOH/KOH)	20-180
Templated Carbon	Aqueous (NaOH/KOH)	120-350
AC	Aqueous (NaOH/KOH)	200-400

The capacitance can be calculated using the below equation

$$C = \frac{A\epsilon_0}{d}$$

where d; the length of the electric double layer (Debye length); ϵ_0; the permittivity of the void; A; the surface area and C; the capacitance.

The super magnetic energy storage system preserves energy in electrical form, represented in Figure 9.14. It does not convert energy into any other form. The system has a large energy density and low response time. It has a long life cycle as well as cycling time [39]. The disadvantages are material loss due to mechanical stress and cost ineffective. The system stores electrical energy in the form of magnetic energy. The superconducting coils in the system are made up of niobium-titan material. It is maintained under a cryogenic system 9-2700 °C). The system has an overall efficiency of 95%. It consists of four main components: the superconducting magnet (placed in vacuum), refrigeration system (commenting under cryogenic temperature), power maintaining system (consisting of transistors, capacitors, inductors etc.), and control system (monitor different variables of the system like pressure and temperature).

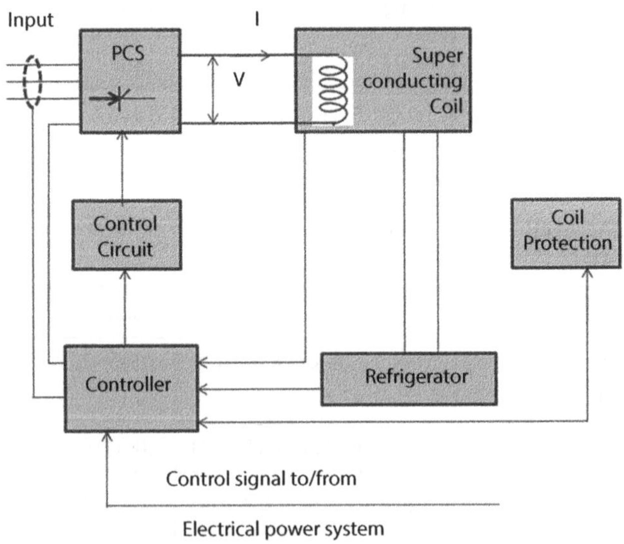

Figure 9.14 A pictorial representation of super magnetic energy storage (SMES) system [40].

If " I" is the flow of current and L is the inductance, then the highest energy storage is

$$E_{m^\infty} = \frac{1}{2} LI^2 \approx \frac{1}{2} B_s INS$$

Figure 9.15 represents the electric circuit of the super magnetic storage system.

9.3.4 The Electrochemical Energy Storage System

The arrangement comprises rechargeable and flow batteries. It conserves electrical energy as chemical energy. It is an ancient yet advanced technology [3]. The stored chemical energy is evolved into electrical energy by using a redox reaction [42]. Its energy density ranges between 10-13 Wh/kg. Its efficiency is 70%. The system requires negligible maintenance [26].

The battery consists of two electrodes; positively charged and negatively charged. The electrodes remain inundated in a solid, liquid, or viscous electrolyte. During the off-peak phase, a redox reaction occurs and anions are released into the electrolyte from the anode. The electrons that remained in the anode after the release of anions travel to negatively charged via an external circuit. Hence the current is generated because of electron flow. During the high-peak state, these electrons move from the cathode to the anode. Since one battery is insufficient for the desired need, many batteries are interconnected in an array to get the required result. The standard reduction potential for various chemicals used in battery energy storage

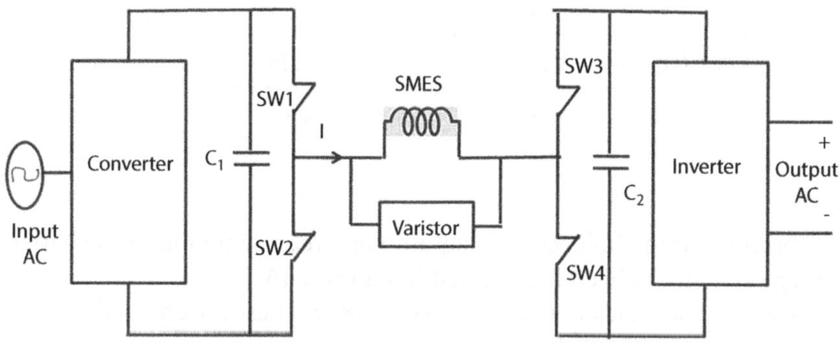

Figure 9.15 Electrical circuit for SMES [41].

Table 9.6 The standard reduction potential of various chemicals in battery energy storage (BES) [3].

Reduction half reaction	Standard electrode potential (V)
Li^+ (aq) + e^- → Li(s)	-3.04
F_2(g) + $2e^-$ → $2F^-$ (aq)	+2.87
$S_2O_82^-$ (aq) + $2e^-$ → $2SO_42^-$ (aq)	+2.01
Na^+ (aq) +e^- → Na (s)	-2.71
O_2 (g) + $4H^+$ (aq) + $4e^-$ → $2H_2O$ (I)	+1.23
Br_2 (I) +$2e^-$ → $2Br^-$ (aq)	+1.09
Mn^{2+} (aq) + $2e^-$ → Mn (s)	-1.18
Ag^+ (aq) + e^- → Ag (s)	+0.80
Zn^{2+} (aq) +$2e^-$ → Zn (s)	-0.76
Fe^{3+} (aq) + e^- → Fe^{2+} (aq)	+0.77
I_2 (I) + $2e^-$ → $2I^+$(aq)	+0.54
Cr^{3+} (aq) + $3e^-$ → Cr (s)	-0.74
Cu^{2+} (aq) + $2e^-$ → Cu (s)	+0.34
Fe^{2+} (aq) +$2e^-$ → Fe (s)	-0.44
Sn^{4+} (aq) +$2e^-$ → Sn^{2+} (aq)	+0.15
S (s) +$2H^+$ (aq) +$2e^-$ → H2S (g)	+0.14
Sn^{2+} (aq) + $2e^-$ → Sn (g)	-0.14
V^{3+} (aq) +e^- → V2+ (aq)	-0.26
$2H^+$ (aq) +$2e^-$ → H2 (g)	0.00

has been listed in Table 9.6. A diagrammatic representation of the battery storage system has been highlighted in Figure 9.16.

Lead-acid batteries were discovered in 1859. These are charged with the reverse flow of current. Lead-acid batteries are consistent and hence they are established worldwide [43, 44]. This battery can discharge a large quantity of energy because of its high discharge rate. It needs fewer repairs, and

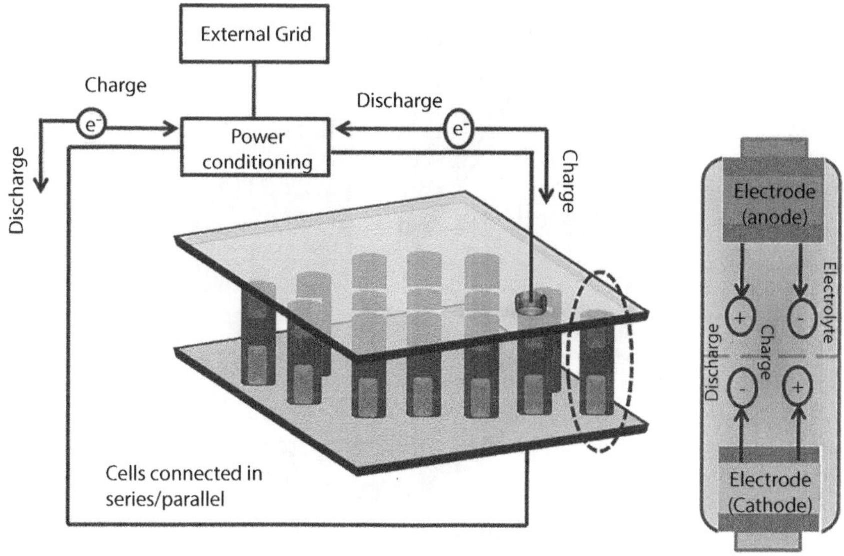

Figure 9.16 A diagrammatic representation of battery energy storage system (BESS) [26].

recharging does not affect its capacity. The recycling rate is 99.4% [45]. It is used in various industries like telecommunication, power system, electrical vehicles, and many more [46]. These batteries comprise a cathode and an anode. The cathode is composed of sponge lead and the anode is made of lead dioxide. The electrodes are submerged in sulphuric acid (38%). The lead acid batteries also contain various other materials as represented in Figure 9.17.

The Pb from the negative pole suspends in an electrolytic solution and forms $PbSo_4$ with $2e^-$ behind during the discharge state. These electrons travel through an external circuit and reach the positive pole. Then, the lead from the positive pole dissolves in the electrolyte. Hence, sulphuric acid is lost and forms water. The poles become alike during battery discharge leading to a voltage drop. The battery is then recharged by electrical current, thereby restoring the chemical difference between electrodes [44]. The principle on which these batteries are based is shown in Figure 9.18. The mechanism of lead acid batteries is represented below in the form of chemical reactions

Negative electrode reaction:

$$Pb(solid) + HSO_4^-(aqueous) \rightarrow PbSO_4\,(solid) + H^+(aqueous) + 2e^-$$

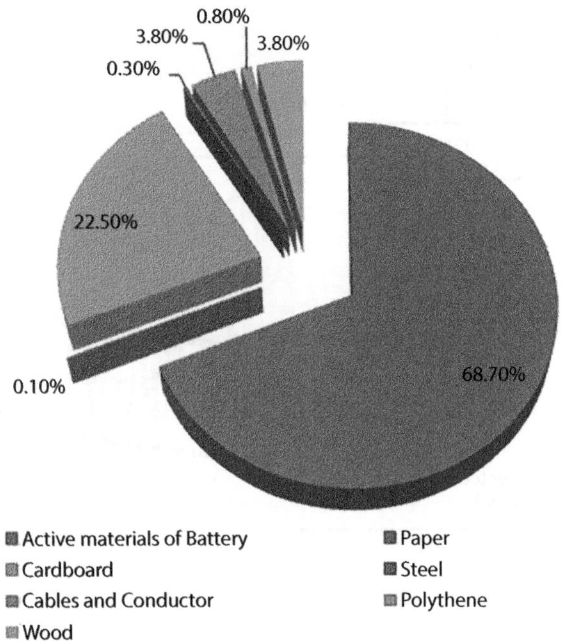

Figure 9.17 Proportion of various materials used in lead acid batteries [47].

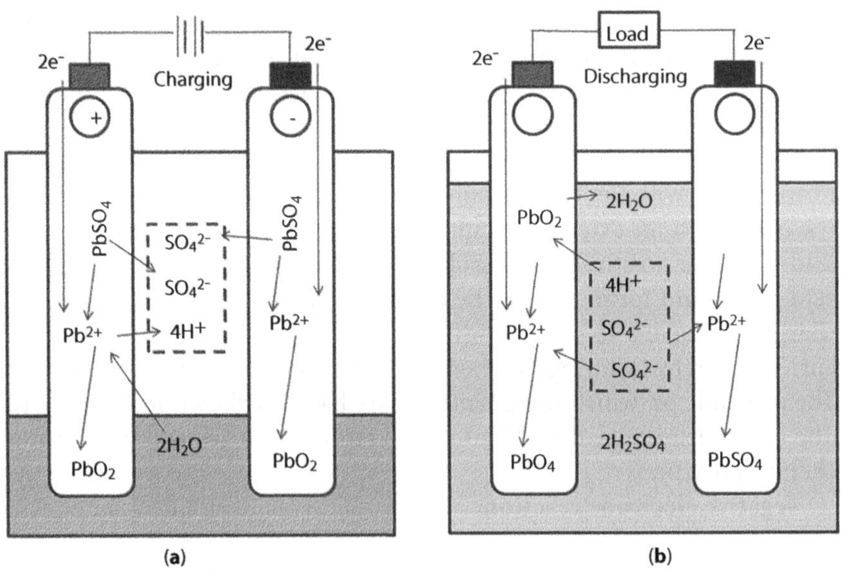

Figure 9.18 Principle of lead acid battery during (a) charging phase and (b) discharging phase [48].

Positive electrode reaction:

$$PbO_2(solid) + HSO_4^-(aqueous) + 3H^+(aqueous) + 2e^- \rightarrow PbSO_4(solid) + 2H_2O(I)$$

Overall Chemical Reaction:

$$Pb(solid) + PbO_2(solid) + 2H_2SO_4(aqueous) \rightarrow 2PbSO_4\ (solid) + 2H_2O(I)$$

Flooded lead-acid batteries are extremely economic and hence widely used [49]. This battery requires a lot of maintenance. Unlike lead acid batteries, the electrolytes can freely move around the battery cell. It has a weak structure and elevated resistance [49]. The diagrammatic representation of the same is shown in Figure 9.19.

The sealed lead-acid battery is the form of a lead-acid battery composed of precipitated H_2So_4 electrolyte. These are partially sealed and contain vents. Gas can escape through these valves during the charging phase. These are safe to use but are costlier. They can be of two types: absorbed

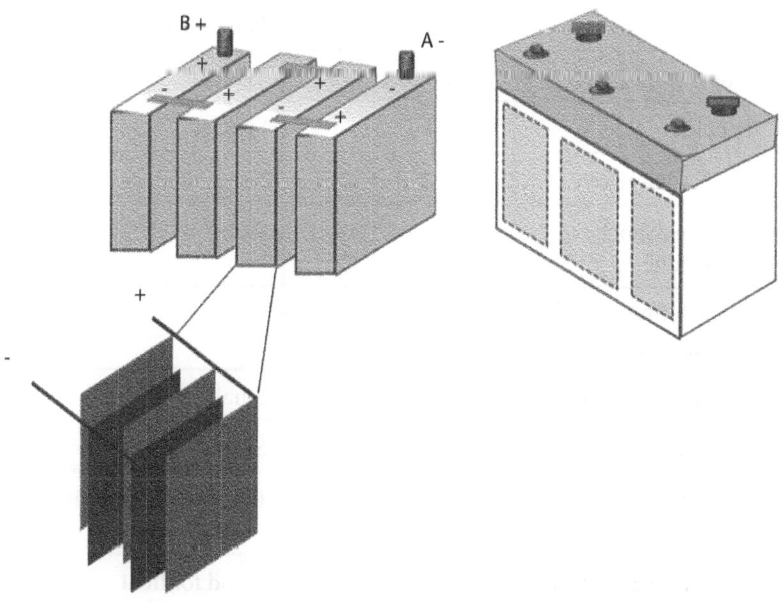

Figure 9.19 Diagrammatic representation of a flooded or wet lead acid cell [80].

glass matte and gel cell batteries. The difference between them has been presented in Table 9.7.

Lithium-metal battery is a disposable battery which uses "Li" as the anode. Lithium is useful as an anode because it is highly reactive and can easily release an electron to form Li^{+}. Lithium is one of the lightest metals and hence lithium batteries are very light. It has wide applications. The main problem associated with lithium metal is its explosive nature. Therefore, researchers are working to solve this issue by developing new cathode material for a surface Kota for the anode.

Lithium-ion batteries are gaining popularity because this type of battery has a prolonged life cycle, high power, and less self-discharge speed; it is lighter in weight and poses high charge density [51, 52]. It consists of two

Table 9.7 The comparision absorbed glass matte (AGM), and gel batteries [50].

Property	Deep cycle AGM	Deep cycle gel
Maintenance	Not Needed	Not Needed
Cost	Low	Highest
Cycle Life	Good	Good
Shipping	Non-Hazardous	Non-Hazardous
Temperature Range	Wide	Wide
Set points	Special	Special
Self-Discharge rate	Low	Low
Vibration/Shock Resistance	Good	Good
Recyclability/ Infrastructure	Good	Good
Discharge currents	Wide range	Small
Charge Currents	Wide Range	Small
Applications	High power/Fast Charge	Low power applications only
Float Characteristics	Excellent float characteristics	Good for float applications

current collectors, an electrolyte, a separator, a cathode and an anode [53] The anode is formed of metal-oxides compounded with lithium. The cathode is formed of a carbon isotope (graphite). The electrolyte constitutes lithium salt suspended in organic carbonate [6]. At the discharge phase, the "Li" present in the positively charged electrode ionized because of variation in electrolytic concentration. The ions are then washed out by an electrolytic substance towards the negatively charged electrode. Once the ions reach to cathode it combines with the peripheral electron and gets converted into a "Li" atom. The flow of "Li^+" generates liberated electrons in the positively charged electrode that travels via a peripheral circuit and reaches the negatively charged electrode to recombine with "Li^+". The divider is used to stop the direct passage of electrons from the positive electrode to the negative electrode. The process is reversible during the high-peak phase. Figure 9.20 represents the mechanism of phases of lithium-ion batteries. The lithium-ion batteries can be of various types, differing in their characteristics. Different lithium-ion batteries are listed in Table 9.8, which also highlights their properties.

Table 9.8 Various Li-ion batteries and their properties [55].

Type of lithium ion (Li-ion) battery	**Lithium iron phosphate**	**Lithium nickel cobalt aluminum oxide**	**Lithium titanate**
Available Since	1996	1999	2008
Short form	LFP or Li-Phosphate	NCA or Li-aluminum	LTO or Li-titanate
Cathode	$LiFePO_4$	$LiNiCoAlO_2$	$LiMnO_2$ or $LiNiMnCoO_2$
Anode	Graphite	Graphite	$Li_4Ti_5O_{12}$ (Titanate)
Voltages	3.2 V, 3.3 V nominal; Typical operating range: 2.5–3.65 V/cell	3.6 V nominal; Typical operating range: 3–4.2 V/cell	2.4 V nominal; Typical operating range: 1.8–2.85 V/cell

(*Continued*)

Table 9.8 Various Li-ion batteries and their properties [55]. (*Continued*)

Type of lithium ion (Li-ion) battery	Lithium iron phosphate	Lithium nickel cobalt aluminum oxide	Lithium titanate
Specific Energy (Capacity)	90–120 Wh/kg	200–260 Wh/kg; 300 Wh/kg predictable	50–80 Wh/kg
Charge (Crate)	1 C typical; charges to 3.65 V; 3 h charge Time typical	0.7 C; charges to 4.2 V (most cells); 3 h Charge typical; fast Charge possible with some cells	1 C typical; 5 C maximum; charges to 2.85 V
Cycle life	1000–2000 (related to depth of discharge, temperature)	500 (related to depth of discharge, temperature)	3000–7000
Thermal runaway	270 °C (518 °F). Very safe battery even if fully charged.	15 °C or 302 °F typical; high charge promotes thermal runaway	One of the safest Li-ion batteries
Cost	~580$/kWh (Source: RWTH Aachen)	~350$/kWh (Source: RWTH Aachen)	~1005$/kWh (Source: RWTH Aachen)
Applications	Portable and Stationary needing high load currents and endurances	Medical devices, industrial, electric power trains (Tesla)	UPS, electric power train (Mitsubishi i, MiEV, Honda Fit EV), solar Powered street lighting

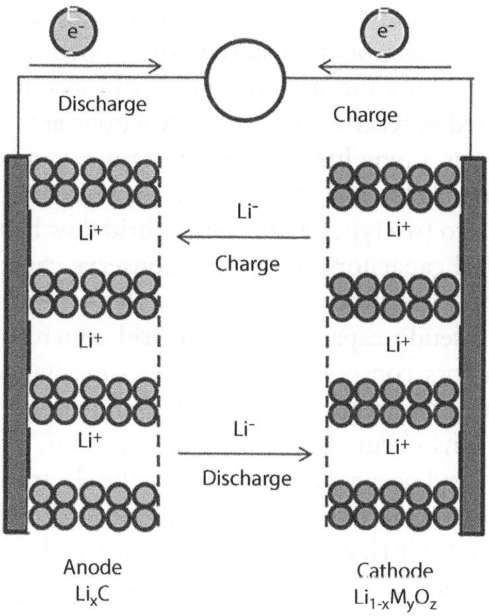

Figure 9.20 The operation of Li-ion battery; [54].

Lithium polymer batteries are similar to Lithium-ion batteries. They use a dry-solid polymeric electrolyte. This form of electrolyte mixes make lithium polymer batteries easier compared to others. They have poor conductivity because of high internal resistance [56]. Nowadays, the gelled electrolyte is added to solve the above-mentioned complication. It has reduced self-discharge rate and a long life cycle. These are slimmer and lighter in weight. Sodium beta batteries use sodium beta alumina as an electrolyte membrane. It carries an ion between the sodium anode and cathode.

A metal air battery is made up of metal electrode as anode like lithium, zinc, magnesium, etc. These are cheaper and more compact [52, 57]. The cathode is a bifunctional air electrode made up of metal mesh or carbon material. The electrolyte could be aqueous or non-aqueous. The batteries are eco-friendly and lighter. They use low-cost metals and have a high charge density. The batteries are difficult to recharge but can be refuelled [6]. Nickel-based batteries use nickel oxide hydroxide as an anode [58]. It is highly stable, with a long life, and has temperature tolerance.

Flow batteries are advanced aqueous electrolyte batteries. The electrolyte consists of more than one electroactive material [3, 6, 59]. This electrolyte converts chemical energy into electrical energy. It does not contain any self-discharge. The extra electrolytic material is conserved in two individual barrels and is passed via the reactor to cause a chemical reaction [3, 6]. The electrolytic materials are gapped through the minute membrane which allows the selective number of ions to pass. These ions take part in the redox reaction. Chemical reactions are reversible and can be recharged. It has a long life, less response time, and is highly efficient. Depending upon the electrochemical materials present in the flow battery, it is categorised into two types: redox and hybrid flow battery [26].

Electrochemical capacitors are used to conserve charge in the form of ions. These devices are based on the principle of Faradic reaction. Two main types are pseudo-capacitors and hybrid supercapacitors. Pseudo-capacitors capacitors conserve charge. Their capacitance depends upon the flow of charge between the electrolyte and pseudo-electrode [60]. The transfer of ions occurs due to faradic reactions. The capacitance is non-electrostatic and occurs on the surface of the electrode. The electrodes are combined with metal collectors that can collect current from electrodes dipped in electrolytes [61]. Pseudo-capacitors are double-layered and form supercapacitors. The charge-storing mechanism in the pseudo-capacitors depends upon the transfer of electrons.

Hybrid supercapacitors are more advantageous compared to pseudo-capacitors. These are based on faradic and non-faradic storage mechanisms. They have high power density, stability, and durability. They use the advantages of pseudo-capacitors and electrochemical capacitors. They have high specific capacitance. The hybrid supercapacitor is classified into Battery, Asymmetric and Composite types depending upon its electrode.

9.3.5 The Chemical Energy Storage System

The chemical energy storage system stores energy as chemical bonds. The reaction between molecules transfers electrons and generates energy. It transforms energy from one chemical form to another form. It can conserve approximately 100GWh [32]. Chemical energy storage systems can be of two forms: a thermo-chemical energy storage system and an electrochemical energy storage system. The electrochemical energy storage system utilizes fuel cells.

The chemical energy photochemical reaction can be strictly converted into electrical energy to generate H_2 and H_2O with the help of fuel cells. The concept was originally developed by Christian Friedrich in 1838 [62]. It stores energy which can generate hydrogen during the electrical degradation of water. Fuel cells are the amalgamation of engines and batteries. The mechanism is following the working mechanism of a battery. It is similar to an engine in a way that it functions till the fuel is present [62]. The fuel cell has a pro-long life of around 15 years. It has more charge and discharge rate. It also causes more energy density. Fuel cells can be applied for narrow and wide-scale applications. The major disadvantages of fuel cells are low efficiency and short life expectancy. Nowadays, hydrogen is gaining attention because it is the cleanest and lightest fuel known so far. Hydrogen can be efficiently conserved and reused whenever necessary [32]. Figure 9.21 provides a diagrammatic representation of the working principle of the fuel cell. During off-peak time water is electrolysed to produce hydrogen. During peak time the produced hydrogen is mixed with oxygen to generate electricity. The fuel cells consist of two electrodes separated by an electrolyte. The electrodes that contain hydrogen fuel act as an anode and that contain oxygen act as a cathode. Hydrogen is degraded into hydrogen ions and electrons at the anode. Do hydrogen ions pass through the electrolyte membrane and reach up to the negative electrode? The formed electrons travel from the positive electrode to the negative electrode via an external circuit to be combined with hydrogen ions and oxygen to produce water.

$$Anode: H_{2^\circ} \Rightarrow 2H^+ + 2e^-$$

$$Cathode: \frac{1}{2}O_2 + 2H^+ + 2e^- \Rightarrow H_2O$$

$$Overall: H_2 + \frac{1}{2}O_2 \Rightarrow H_2O$$

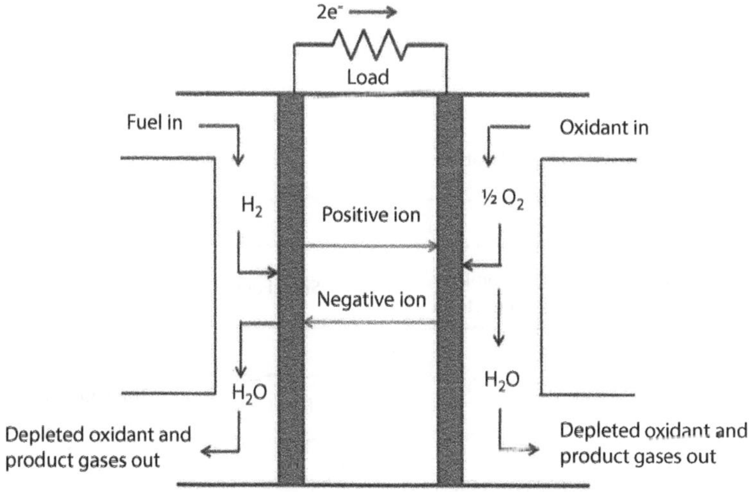

Figure 9.21 A diagrammatic representation of the working principle of fuel cell [62].

Depending upon the type of fuel and electrolytes used, the fuel cells are categorized into eight ways.

Solar fuels utilize solar energy and convert it to alternative fuel energy. Concentrators are used to capture more temperature from low sunlight and receivables extract radiation energy from it [63–65]. The produced high temperature is used to generate alternative fuels that can be conserved and transported [66]. Figure 9.22 summarises the manufacturing details of solar fuel cells.

The supremacy of the system is its 100% efficiency and less energy loss. It has a high capacity to run heat engines. Solar future is economic.

Solar hydrogen generates hydrogen by utilising solar energy. The combustion of hydrogen produces water as a by-product. Solar hydrogen can be produced in five different ways; the same has been represented in Figure 9.23.

Water can be used to produce hydrogen by the process of thermo-analysis. Fossil fuels are used to produce hydrogen by solar cracking. The combination of water and fossil fuels is utilised to produce hydrogen bad mechanism of solar reforming and solar gasification. The formed hydrogen can easily be used as fuel to transform its chemical energy into electrical energy.

Due to the difficulty of preservation and transportation of hydrogen, 16 hydrocarbons are used. The major disadvantage of hydrocarbons is

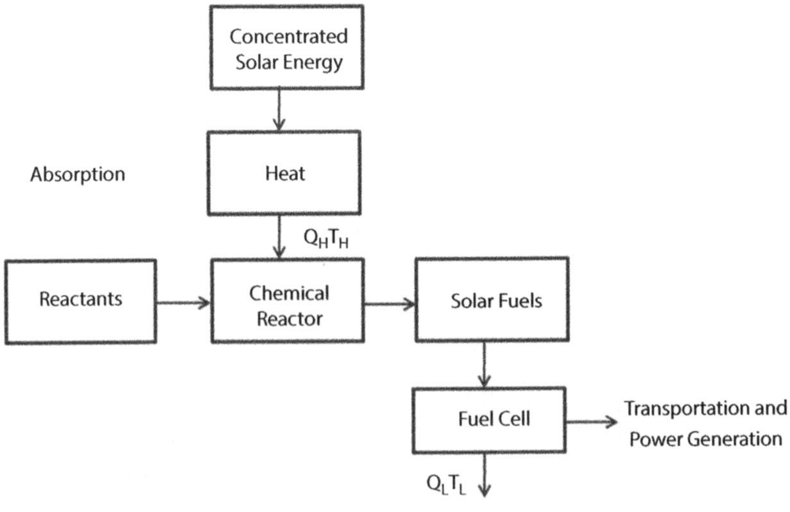

Figure 9.22 Highlights of manufacturing system of solar fuel cells [67].

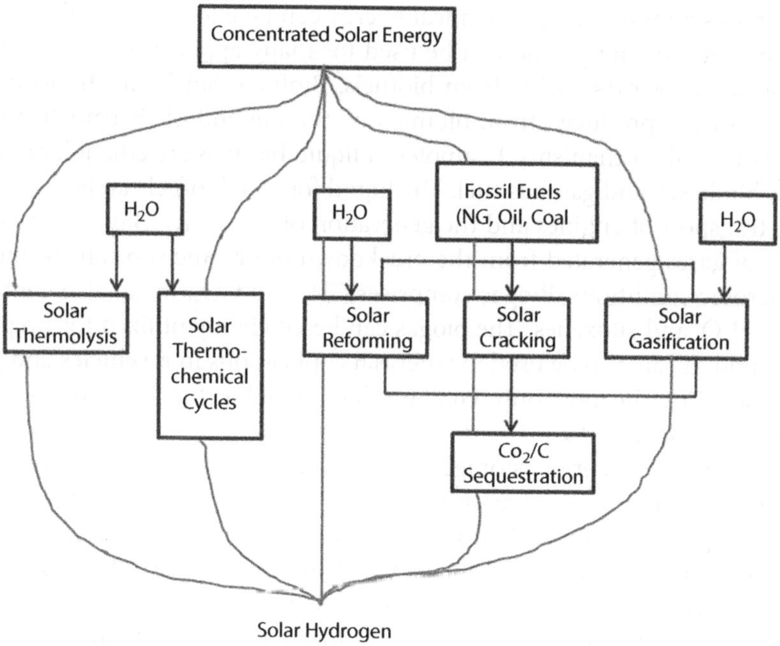

Figure 9.23 Five various ways of producing solar hydrogen from concentrated solar power (CSP) [6].

the release of carbon dioxide. Therefore, researchers use carbon-neutral hydrocarbons.

Solar metals can be easily stored and transported. They can generate high energy at high temperatures due to combustion. The process results in the production of metal oxides which can be reused and reduced. Metals can easily be extracted from their oxide form by the process of electrolysis and heat. But it can consume a greater amount of energy and cause more pollution. Hence, the metal extraction process is restored by high heat energy generated from solar energy [68].

Solar chemical heat pipe utilizes heat energy from the sun for reversible endothermic chemical reactions. The products obtained from the reactions can be conserved and transported easily. The reversible exothermic reactions are used to extract the heat energy stored in the products. The main advantage of the system is the generation of reactants after extracting heat energy from the products [69, 70].

Biofuels are an environmentally friendly, easily available, and reliable source of energy. Biofuels are mainly obtained from plants. The plants naturally trap solar energy and store it as chemical energy; the process is known as photosynthesis. Stored chemical energy can be effortlessly evolved into other states of energy and can be used for many applications. Three percent of the world's fuel is from biofuels. Biofuels can be in the form of liquid or gas produced from biomass by the method of thermochemical or biological mechanisms. Examples of liquid biofuels are ethanol, methanol, biodiesel, and gaseous fuel. The liquid form of biofuel can be used for the transport of engines and the generation of electricity. Biogas is a mixture of gases generated from the breakdown of organic substances under anaerobic conditions. Biogas comprises CH_4 and CO_2. It can also contain H_2S, H_2O, and siloxanes. The biogas can be efficiently utilized for preparing food. It can also be used for thermal applications in e-vehicles and gas engines [71]. The intercontinental production of bio-fuel in 2018 has been presented in Table 9.9.

Aluminium energy storage utilizes aluminium to store energy. Aluminium is inexpensive and readily available in nature. During the charging phase Al_2O_3 and $Al(OH)_3$ are used to cast-off aluminium as an element after reduction. The formed aluminium can simply be conserved and moved from one place to another whenever necessary. During the discharge phase, the aluminium is oxidized to derive hydrogen, heat, and Al_2O_3 or $Al(OH)_3$. The H_2 and the produced heat which are the by-products can be utilised for the production of electricity and thermal usage, respectively.

A Hybrid Energy Storage System

It is the system that utilizes the advantages of two or more energy storage systems. The hybrid system can function in either of two modes: high power and high energy. High-power energy storage systems respond

Table 9.9 The intercontinental production of bio-fuel in 2018 [71].

Country	Production (thousand barrels/day)	Share in the world's total biofuel production in 2018 (%)
Argentina	70.6	2.7
Brazil	693.2	26.5
China	68	2.6
Germany	75.8	2.9
USA	1190.2	45.5

fast and include flywheel energy storage, and various electrical energy storage systems. A high-energy energy storage system includes fuel cells electrical energy system, chemical energy system and many more. Tables 9.10, 9.11 and 9.12 show various parameters for different energy storage systems.

Table 9.10 The characteristics of different energy storage systems—I [26].

Technology	Energy density (Wh/L)	Power density (W/L)	Specific energy (Wh/Kg)	Specific power (W/Kg)	Rated energy capacity (MWh)	Power rating (MW)
Liquid air Storage	4-6 times than CAES at 200 bar	-	214	-	2.5	0.3-200
PHS	0.5-2	0.5-1.5	0.5-1.5	-	180-8000	30-5000
TES	80-500	-	80-250	10-30	-	0.1-300
Large Scale CAES	2-6	0.5-2	30-60	-	580-2860	110-1000
Hydrogen Fuel Cells	500-3000	500+	800-10,000	5-800	0.312-39	<58.5
Overground Small CAES	Higher than large scale CAES	Higher than large scale CAES	140-300 bar	-	0.0083-0.01	0.003-10
Solar Fuel	500-10,000	-	800-100,000	-	-	0-20
Flywheel	20-80	1000-5000	5-100	400-1500	0.0052-5	0.25-20
SMES	0.2-6	1000-4000; 2500	0.5-5; 10-75	500-2000	0.0008-0.015	0.1-10
Lead Acid	50-80	10-400	25-50	75-300	0.001-40	0-40
Supercapacitor	10-30	100,000+	0.05-15	500-10,000	0.0005	0.0-3
Li-ion	200-500	1500-10,000	75-200	150-2000	0.004-10	0-100
Capacitor	0.05-10	100,000+	0.05-5; <0.05	100,000>300-107	-	0-0.005
Na-S	150-300	140-180	100-240	90-230	0.4-244.8	<34
PSB	20-30	<2	15-30	-	Upto 120	0.004-15
NiCd	15-150	80-600	45-80	150-300	6.75	0-40
ZnBr	30-65	<25	30-80	100	0.05-4	0.05-10
VRB	16-35	<2	10-30	166	<60	0.03-50

Table 9.11 The characteristics of different energy storage systems—II [26].

Technology	Daily self-discharge	Lifetime (years)	Cycling times	Discharge efficiency	Cycle efficiency	Response time
Liquid Air Storage	Small	25+	-	-	55-80	Minutes
PHS	Very Small	40-60	10,000-30,000	87	70-87	Minutes
TES	>1	20-30	-	-	30-60	Minutes
Large Scale CAES	Very Small	20-40	8000-12,000	70-79	42-54; AA-CAES:70	Minutes
Hydrogen fuel Cell	Almost Zero	20	20,000	59	20-66	Seconds
Over ground small CAES	Very small	23+	30,000	75-90	-	Seconds-Minutes
Solar fuel	Almost Zero	-	-	-	20-50	-
Flywheel	> 20 % per hour	15-20	20,000	90-93	90-95	Seconds
SMES	10-15	20-30	100,000+	95	95-98	Milli-seconds
Lead acid	>0.3	5-15	21,000	85	63-90	Milli-Seconds
Super-capacitor	20-30	10-30	100,000+	95-98	84-97	Milli-seconds
Li-ion	>1	14-16	200-1800	85	75-97	Milli-Seconds
Capacitor	50 in 15 min.	1-10	5-50,000	75-90	60-70	Milli-seconds
Na-S	Almost Zero	10-20	2500-4500	85	75-90	Milli-Seconds
PSB	Almost Zero	10-15	-	-	60-75	20 Milli-seconds
NiCd	>0.6	15-20	2000-3500	85	60-83	Milli-Seconds
ZnBr	Small	5-10	1500-2000	60-70	65-80	>1/4 Cycle
VRB	Very Small	10-20	12,000	75-82	65-85	>1/4 Cycle

Table 9.12 The characteristics of different energy storage systems—III [26].

Technology	Suitable storage duration	Discharge time at related power	Power capital cost ($/Kw)	Energy capital cost ($/Kwh)	Operating and Maintenance cost	Maturity
PHS	Hours-Months; Longterm	1-24h	2000-4300	5-100	~3$/kW/year; 0.004 $/Kwh	Mature
Large Scale CAES	Hours-Months; Longterm	1-24h	400-1000	2-120	19-25 $/kW/ year; 0.003 $/ Kwh	Developed
Overground small CAES	Hours-Months; Longterm	30s-40m; 3h	517-1550	200-250	Very low	Developed
Flywheel	Seconds-Minutes; Shortterm	8s-15m	250-350	1000-14000	~20 $/kW/ year; 0.004 $/ Kwh	Commercialized
Lead acid	Minutes-Days; Short to medium term	Seconds-Hours	200-600	50-400	50$/ kW/year	Mature
Li-ion	Minutes-Days; Short to medium term	Minutes-hours	900-4000	600-3800	-	Commercialized
Na-S	Long Term	Seconds-hours	350-3000	300-500	~80$/kW/year	Commercialized
NiCd	Minutes-Days; Short and long term	Seconds-hours	500-1500	400-2400	~20$/kW/year	Mature
VRB	Hours-Months; Longterm	Seconds-hours	600-1500	150-1000	~ 70$/kW/year	Developing
ZnBr	Hours-Months; Longterm	Seconds-hours	200-2500	150-1000	-	Demonstration
PSB	Hours-Months; Longterm	Seconds-hours	700-2500	150-1000	-	Developing
Capacitor	Seconds-hours	Milliseconds - 1h	200-400	500-1000	~13 $/kW/ year; < 0.05 $/Kwh	Commercialized
Supercapacitor	Seconds-hours; Short term	Milliseconds - 1h	100-450	300-2000	< 0.005 $/Kwh; ~ 6 $/kW/ year	Developing
SMES	Minutes-Hours; Sort term	Milliseconds-30min.	200-489	500-72000	18.5 $/kW/ year; 0.001 $/ Kwh	Early Commercialized

(*Continued*)

Table 9.12 The characteristics of different energy storage systems—III [26]. (*Continued*)

Technology	Suitable storage duration	Discharge time at related power	Power capital cost ($/Kw)	Energy capital cost ($/Kwh)	Operating and maintenance cost	Maturity
Solar Fuel	Hours-Months	1-24 h	-	-	-	Developing
Hydrogen FuelCell	Hours-Months	Seconds-24h	500-3000	2-15	0.0019-0.0153 $/kW	Developing
TES	Minutes-Months	1-24h	100-400	3-60	-	Early Commercialized
Liquid Air Storage	Long Term	Several Hours	900-1900	260-530	-	Developing

9.4 Implementations of Energy Storage Systems

Earlier energy storage systems were used only for power supply. Nowadays, energy storage systems can also be utilized for various other usages like load levelling, peak shaving, and many more. These are also used in the field of telecommunication, transportation, satellite working, etc.

9.5 Commercial Prototype of Energy Storage Systems

Energy storage systems are an important part of the power system. Its various functions are energy storage and transportation, frequency regulation, power system stability, etc. [72]. Renewable sources of energy have also been used extensively in recent years and are replacing traditional power plants. Energy storage systems are creating new commercialization by linking consumers and producers [72]. Due to the advancement of technology maintenance and operation costs are reduced.

The profit for the specific business depends on the specific energy storage system. Policies to support this business are also needed.

9.6 Environmental Repercussions of Energy Storage Systems

Energy storage systems are environmentally friendly and release fewer greenhouse gases, thereby protecting our environment. These systems help to reduce environmental pollution, which is a great concern in today's world. The increasing demand for energy increases the global carbon dioxide emission by 1.7%. Compared to 2010, 70% of emissions increased in 2013. The excessive use of coal also leads to an increase in the carbon dioxide content of the environment. The countries which exhibit high demand of energy are the USA, India, and China. France, Germany, Japan, and the UK have reduced their emissions.

The high energy demand resulted in a change in climatic conditions. From 2017 to 2018, the emissions rose by 0.5%. The use of renewable and nuclear energy helped to reduce emissions by 25% as recorded in 2018.

Energy systems are cost-effective and help to fight climatic changes. The increased capacity of energy storage systems removes the need for more generating units and transporting and distributing systems [73].

9.7 Energy Storage Guidelines

The energy storage system earned popularity due to its multi-functional nature. These are reliable, durable, and eco-friendly. Worldwide energy storage usage doubled from 2017 to 2018. Appropriate policy and regulatory authorities are needed for the proper usage and maintenance of these energy storage systems. Policymakers should procure laws to create an appropriate market and encourage investors to invest in energy storage systems.

9.8 Blockades and Effective Solutions

Though energy storage systems possess various applications and advantages, they also have various barriers. These barriers, which hinder their usage in a widespread manner, can be categorised in several different ways: developer business model barrier, cross-cutting barrier, economic barrier, technological barrier, regulatory barrier and utility barrier. Every hurdle needs to be tackled for the global application of energy storage systems.

9.9 Future Prospects

Currently, the application of conventional energy sources has increased and is expected to increase in future. The expected global capacity of photovoltaic cells is 3142 GW. It will be the major source of energy by the next

20 years. The use of wind energy will also increase, from 5% to 13% by the year 2040. The combined capability of solar and wind energy will be ~ 5000 GW. Taking into account the other conventional energies, it is expected to increase its share from 26% to 41% by the year 2040 [74].

The large-scale usage of energy sources is increasing day by day. According to the prediction by Bloomberg NEF, the cosmopolitan potency of stored energy will enhance from ~15 GWh to 3000 GWh by the year 2040 [75].

9.10 Conclusion

The present chapter discussed the advantages, disadvantages, and usage of various energy storage systems. With the increase in the need for renewable energy sources, the globe is moving towards energy storage systems. The comparison between various energy storage systems has also been highlighted. A proper energy storage system and better policies for its handling are needed. A proper understanding of these energy storage systems is essential for their proper utilization. Hence, the chapter dealt with every possible aspect of various energy storage systems needed for their widespread application.

References

1. *Global Energy and CO_2 Status Report*, Energy Demand, International Energy Agency (IEA), Paris, France; 2019, p. 29.
2. GLOBE-Net. https://globe-net.com/global-energy-demand-rose-by-2-3-in-2018-its-fastest-pacein-the-last-decade/ (accessed on 5 February 2020).
3. Nadeem, F., Hussain, S.M.S., Tiwari, P.K., Goswami, A.K., Ustun, T.S., Comparative review of energy storage systems, their roles, and impacts on future power systems. *IEEE Access*, *7*, 4555–4585, **2019.**
4. U.N. Environment and International Energy Agency (2017): Towards a zero-emission, efficient, and resilient buildings and construction sector, Global Status Report, 2017.
5. Hossain, E., Murtaugh, D., Mody, J., Faruque, H.M.R., Haque Sunny, M.S., Mohammad, N. A., Comprehensive review on second-life batteries: Current state, manufacturing considerations, applications, impacts, barriers & potential solutions, business strategies, and policies. *IEEE Access*, *7*, 73215–73252, **2019.**
6. Chen, H., Cong, T.N., Yang, W., Tan, C., Li, Y., Ding, Y., Progress in electrical energy storage system: A critical review. *Prog. Nat. Sci.*, *19*, 291–312, **2009.**

7. Poullikkas, A., A comparative overview of large-scale battery systems for electricity storage. *Renew. Sustain. Energy Rev., 27*, 778–788, **2013**.
8. Whittingham, M.S., History, evolution, and future status of energy storage. *Proc. IEEE, 100*, 1518–1534, **2012**.
9. The History and Development of Batteries. https://phys.org/news/2015-04-historybatteries.html (accessed on 23 March 2020).
10. History of Biofuels - BioFuel Information. http://biofuel.org.uk/history-of-biofuels.html (accessed on 23 March 2020).
11. Fuel Cell History, *Fuel Cell Today*. http://www.fuelcelltoday.com/history (accessed on 24 March 2020).
12. Lead-Based Batteries Information. Battery University. https://batteryuniversity.com/learn/article/lead_based_batteries (accessed on 19 June 2020).
13. Hydropower, Explained. https://www.nationalgeographic.com/environment/globalwarming/hydropower/ (accessed on 19 June 2020).
14. History of Flywheel Energy Storage Systems | Gerotor AG; Gerotor GmbH Akt. Energiemanagement. http://gerotor.tech/history-of-flywheel-energy-storage-systems/ (accessed on 6 February 2020).
15. A Brief History of Flow Batteries. https://www.upsbatterycenter.com/blog/flow-batteries-history/ (accessed 24 March 2020).
16. Nickel-Based Batteries Information. Battery University. https://batteryuniversity.com/learn/article/nickel_based_batteries (accessed on 6 December 2019).
17. Omar, N., Firouz, Y., Abdel, M., Samba, A., Gualous, H., Coosemans, T.C., den Bossche, P., Mierlo, J., Analysis of nickel-based battery technologies for hybrid and electric vehicles, in: *Reference Module in Chemistry, Molecular Sciences and Chemical Engineering*, Waltham (Ed.), Elsevier. 2014
18. Nickel Iron Battery Information. http://www.nickel-iron-battery.com/ (accessed on 13 March 2020).
19. Sparacino, A.R., Reed, G.F., Kerestes, R.J., Grainger, B.M., Smith, Z.T., Survey of battery energy storage systems and modeling techniques, in *Proceedings of the 2012 IEEE Power and Energy Society General Meeting, San Diego, CA, USA*, 2012, pp. 1–8.
20. Hartenbach, A., Bayer, M., Dustmann, C.-H., The sodium metal halide (ZEBRA) battery, in: *Molten Salts Chemistry*, Lantelme, F., Groult, H. (Ed.), Elsevier, 2013, pp. 439–450.
21. Hassenzahl, W.V., Superconducting magnetic energy storage. *Proc. IEEE*, **1983**, *71*, 1089–1098.
22. O'Donnell, M., Smithrick, J., *Nickel Hydrogen Batteries – an Overview*, NASA Lewis Research Center: Cleveland, OH, USA, p. 13, 1995.
23. Supercapacitor Information. Battery University. https://batteryuniversity.com/learn/article/whats_the_role_of_the_supercapacitor (accessed on 19 June 2020).
24. Jimbo Compressed Air Energy Storage. https://www.thegreenage.co.uk/tech/compressed-air-energy-storage/ (accessed on 19 June 2020).

25. DNK Power. https://www.dnkpower.com/lithium-polymer-battery-guide/ (accessed 24 March 2020).
26. Luo, X., Wang, J., Dooner, M., Clarke, J., Overview of current development in electrical energy storage technologies and the application potential in power system operation. *Appl. Energy*, *137*, 511–536, **2015**.
27. Breeze, P., Power system energy storage technologies, in: *Power Generation Technologies*, P. Breeze (Ed.), Elsevier, 2019, pp. 219–249.
28. International Energy Agency. https://www.ourenergypolicy.org/wp-content/uploads/2017/05/TrackingCleanEnergyProgress2017.pdf (accessed on 30 November 2019).
29. U.S. Department of Energy. Office of Electricity Delivery and Energy Reliability. https://www.energy.gov/sites/prod/files/oeprod/DocumentsandMedia/OE_Energy_Storage_Program_Plan_Feburary_2011v3.pdf (accessed on 30 November 2019).
30. Antal, A., *Pumped Storage Hydropower: A Technical Review*. Boulder: Department of Civil Engineering University of Colorado Denver (2014).
31. Wang, J., Lu, K., Ma, L., Wang, J., Dooner, M., Miao, S., Li, J., Wang, D., Overview of compressed air energy storage and technology development. *Energies*, *10*, 991, **2017**.
32. Kalaiselvam, S., Parameshwaran, R., Energy storage, in: *Thermal Energy Storage Technologies for Sustainability*, Kalaiselvam, S., Parameshwaran, R., Elsevier, 2014, pp. 21–56.
33. Hebner, R., Beno, J., Walls, A., Flywheel batteries come around again. *IEEE Spectr.*, *39*, 46–51, **2002**.
34. Amiryar, M., Pullen, K., A review of flywheel energy storage system technologies and their applications. *Appl. Sci.*, *7*, 286, **2017**.
35. Alzola, R., Sebastián, R., Quesada, J., Colmenar, A., Review of flywheel based energy storage systems, in *Proceedings of the 2011 International Conference on Power Engineering, Energy and Electrical Drives, Spain*, 2011, pp. 1–6.
36. Sarbu, I., Sebarchievici, C., A comprehensive review of thermal energy storage. *Sustainability*, *10*, 191, **2018**.
37. Muzaffar, A., Ahamed, M.B., Deshmukh, K., Thirumalai, J., A review on recent advances in hybrid supercapacitors: Design, fabrication and applications. *Renew. Sustain. Energy Rev.*, *101*, 123–145, **2019**.
38. Barrade, P., Serge, P., Alfred, R., Series connection of supercapacitors, with an active device for equalizing the voltages, *PCIM Power Conversion and Intelligent Motion*, Nürnberg, Germany, 2000.
39. Zimmermann, A.W., Yang, Y., Young, D.E.A., *R;eview of the State of the Art Superconducting Magnetic Energy Storage (SMES), in Renewable/Distributed Energy Systems*; CPE 610 Mini Project Final Report, EPSRC Centre for Doctoral Training, U.K., 2017, p. 13.
40. Zaman, A. Abdullah, M., Ahmed, S., Monira, N.J., An overview of superconducting magnetic energy storage (SMES) and its applications, in

International Conference on Nanotechnology and Condensed Matter Physics, Dhaka, Bangladesh, 2018.

41. Kumar, N., *Superconducting Magnetic Energy Storage (SMES) System*, Technical Report, p. 5.
42. Electrochemical Energy Storage, Energy Storage - Technologies and Applications, https://www.intechopen.com/books/energy-storage-technologies-andapplications/electrochemical_energy_storage
43. Lach, J., Wróbel, K., Wróbel, J., Podsadni, P., Czerwiński, A., Applications of carbon in lead-acid batteries: A review. *J. Solid State Electrochem.*, *23*, 693–705, **2019**.
44. May, G.J., Lead batteries for utility energy storage: A review. *J. Energy Storage*, *15*, 145–157, **2018**.
45. Study Finds Nearly 100 Percent Recycling Rate for Lead Batteries. Recycling Today. https://www.recyclingtoday.com/article/battery-council-international-lead-battery-recycling/ (accessed on 3 December 2019).
46. Lead Acid Batteries | Pulsetech Products Corporation. https://www.pulsetech.net/support/lead-acid-batteries.html/ (accessed on 3 December 2019).
47. Roy, P., Vats, V.K., Chakrobarty, P.,.Design and Analysis of Maintenance Free Lead Acid Battery System Used in UPS, 2018.
48. Hannan, M.A., Hoque, M.M., Mohamed, A., Ayob, A., Review of energy storage systems for electric vehicle applications: Issues and challenges. *Renew. Sustain. Energy Rev.*, *69*, 771–789, **2017**.
49. The Differences between AGM, GEL and FLOODED Batteries. Battery Guys. https://batteryguys.com/pages/the-differences-between-agm-gel-and-flooded-batteries (accessed on 5 December 2019).
50. Best RV Deep Cycle Battery 2019—Best RV Battery Guide EVER. *Green Living Blog*, https://mozaw.com/best-rv-deep-cycle-battery/ (accessed 5 December 2019).
51. Mu, D., Jiang, J., Zhang, C., Online semiparametric identification of lithium-ion batteries using the wavelet-based partially linear battery model. *Energies*, *6*, 2583–2604, **2013**.
52. Kamali, S. Emergence of energy storage technologies as the solution for reliable operation of smart power systems—A review. *Renew. Sustain. Energy Rev.*, *25*, 135–365, **2013**.
53. Okou, R., Sebitosi, A.B., Pillay, P., Flywheel rotor manufacture for rural energy storage in sub-Saharan Africa. *Energy*, *36*, 6138–6145, **2011**.
54. Barnes, F.S., *Large Energy Storage Systems Handbook*, 1st ed., CRC Press, Florida, 2011.
55. Types of Lithium-Ion Batteries. Battery University. https://batteryuniversity.com/index.php/learn/article/types_of_lithium_ion (accessed on 5 December 2019).
56. Advantages & Limitations of the Lithium-Ion Battery. Battery University. https://batteryuniversity.com/learn/archive/is_lithium_ion_the_ideal_battery/ (accessed on 7 February 2020).

57. Clark, S., Latz, A., Horstmann, B., A review of model-based design tools for metal-air batteries. *Batteries*, *4*, 5, **2018**.
58. Huang, Z., Du, G., Nickel-based batteries for medium- and large-scale energy storage, in: *Advances in Batteries for Medium and Large-Scale Energy Storage*, Menictas C., Kazacos, M.S., Lim, T.M. (Ed.), Elsevier, 2015, pp. 73–90.
59. BU-210b: How does the Flow Battery Work? Battery University. https://batteryuniversity.com/learn/article/bu_210b_flow_battery (accessed on 7 February 2020).
60. Zhao, Y., Zheng, M., Cao, J., Ke, X., Liu, J., Chen, Y., Tao, J., Easy synthesis of ordered meso/macroporous carbon monolith for use as electrode in electrochemical capacitors. *Mater. Lett.*, *62*, 548–551, **2008**.
61. Fang, Y., Liu, J., Yu, D.J., Wicksted, J.P., Kalkan, K., Topal, C.O., Flanders, B.N., Wu, J., Li, J., Self-supported supercapacitor membranes: Polypyrrole-coated carbon nanotube networks enabled by pulsed electrodeposition. *J. Power Sources*, *195*, 674–679, **2010**.
62. Kirubakaran, A., Jain, S., Nema, R.K., A review on fuel cell technologies and power electronic interface. *Renew. Sustain. Energy Rev.*, *13*, 2430–2440, **2009**.
63. Tuller, H.L., Solar to fuels conversion technologies: A perspective. *Mater. Renew. Sustain. Energy*, *6*, 3, **2017**.
64. Steinfeld, A., Kuhn, P., Reller, A., Palumbo, R., Murray, J., Tamaura, Y., Solar-processed metals as clean energy carriers and water-splitters. *Int. J. Hydrogen Energy*, *23*, 767–774, **1998**.
65. Steinfeld, A., Solar hydrogen production via a two-step water-splitting thermochemical cycle based on Zn/ZnO redox reactions. *Int. J. Hydrogen Energy*, *27*, 611–619, **2002**.
66. Steinfeld, A., Solar thermochemical production of hydrogen—A review. *Sol. Energy*, *78*, 603–615, **2005**.
67. Steinfeld, A., Palumbo, R., Solar thermochemical process technology, in: *Encyclopedia of Physical Science and Technology*, Meyers, R.A., Elsevier, Amsterdam, Netherlands: 2003; pp. 237–256.
68. Schunk, L.O., Haeberling, P., Wepf, S., Wuillemin, D., Meier, A., Steinfeld, A., A receiver-reactor for the solar thermal dissociation of zinc oxide. *J. Sol. Energy Eng.*, *130*, 02100, **2008.**
69. Levy, M., Levitan, R., Rosin, H., Rubin, R., Solar energy storage via a closed-loop chemical heat pipe. *Sol. Energy*, *50*, 179–189, **1993**.
70. Luzzi, A., Endothermic reactors for an ammonia based thermo-chemical solar energy storage and transport system. *Sol. Energy*, *56*, 361–371, **1996**.
71. Top five countries for biofuel production across the globe. *NS Energy*, 15 November 2019.
72. Zhuang, X., Huang, R., Liang, C., Rabczuk, T., A coupled thermo-hydro-mechanical model of jointed hard rock for compressed air energy storage. *Math. Probl. Eng.*, 9, 179169, **2014**.
73. US EPA. Electricity Storage. https://www.epa.gov/energy/electricity-storage (accessed on 9 February 2020).

74. IEA World Energy Outlook: Solar Capacity Surges Past Coal and Gas by 2040. https://www.powermag.com/iea-world-energyoutlook-solar-capacity-surges-past-coal-and-gas-by-2040/ (accessed on 12 February 2020).
75. Bloomberg NEF. https://about.bnef.com/blog/energy-storage-investments-boom-battery-costs-halve-next-decade/ (accessed on 12 February 2020).

Index

1-Ethyl-1-methylpyrrolidinium Bromide [C2MPyrr]Br, 87

Acetate ions, 84, 85
Air compression, 147
Amphoteric exchange, 22
Anion Exchange Membranes, 143
Anolyte, 53-58, 63, 171

Batteries, 201
Battery, 51, 53-60, 62-68, 71-75
Bi-ionic potential, 21
Bipolar exchange, 22
Bipolar Plates, 145
Bipolar redox-active molecules, 33
Br3- and Br5-., 88
Bromide batteries, 91
Buffering agent, 84

Capacitor, 4
Capacitors, 215
Carbon nanofiber, 135
Carbon nanotubes, 137
Carbon paper, 135
Catholyte, 53-55, 171
Cation Exchange Membrane, 8, 141
Ce4+/Ce3+ redox couple, 119
Cell circuit, 28
Cetyltrimethylammonium bromide (CTAB), 87, 89
Charge transfer process, 85
Chemical Stability, 13
Chemical vapor deposition technique, 136
Chloride ions, 85
Coating, 29
Compressed air energy storage, 206
Conducting polymer, 84
Conductivity, 5
Copolymerization, 23
Corrosion, 84
Corrosive environment, 31
Cost-effective, 20
Coulombic efficiencies, 17
Coulombic efficiency, 88, 137
Critical current, 83, 85
Cross-contamination, 26
Cross-linked linear polymer, 21
Cryogenic energy storage, 213
Current densities, 135
Current density, 7, 84, 85, 86, 87
Current efficiency, 84
Cycling Performance, 13

Dendrite, 85, 86, 87, 88, 89
Dendrites, 99
Dense Membranes, 14
Depth of self-discharge, 84
Diffusion-limiting current, 85
Discharge capacity, 139
Discharge operation, 131
Donnan effect, 35
Durability, 33

Electric double-layered capacitors, 216
Electric Vehicles, 102
Electro crystallization, 85, 86
Electro-Chemical Energy Storage, 92

Electro-neutrality, 84
Electrocatalyst, 99
Electrochemical capacitor, 228
Electrochemical energy, 177
Electrochemical Impedance
Spectroscopy, 12
Electrochemical properties, 24
Electrochemistry, 98
Electrode surface deposition, 134
Electrodes, 132
Electrolyte dissolution technique, 138
Electrolytes, 3, 137
Electrospinning technique, 133
Electrostatic energy storage, 215
Energy, 2
Energy efficiency, 145
Energy storage system, 104
Energy storage technology, 107-108
Exchange capacity, 140

Fick's diffusion law, 12
Fine crystal grains, 85
Flexible, 3
Flooded lead-acid batteries, 223
Flow batteries, 227
Fly-wheel, 147
Flywheel energy storage, 210
Fossil fuels, 128
Fuel cells, 228

Grafting, 23
Graphite felt, 133

Homogenized electrolyte, 85
Hybrid Flow Batteries, 97
Hybrid supercapacitors, 228
Hydro pumping storage, 147
Hydrocarbon, 2
Hydrogen evolution, 84
Hydrogen evolution reaction, 85
Hydrophilic, 141
Hydrophilic and hydrophobic
domains, 23
Hydrothermal method, 135

Immersion, 13
Impermeable, 20
Indium oxide, 87
International Energy Agency, 128
Ion-selective membranes, 84
Ionic Selectivity, 11

Kinetic polarization, 133

Layer by layer, 29
Lead-acid batteries, 146
Lithium polymer Batteries, 227
Lithium-ion batteries, 91, 224
Lithium-ion battery, 146
Lithium-metal battery, 224

Mechanical stability, 10
Mechanical strength, 24
Membrane, 140
Membranes, 1
Metal air battery, 227
Methansulfonic acid, 87
Micro-porous membrane, 88
Morphology, 34

Nafion membrane, 88, 141
Nafion membranes, 88
Nanofiltration, 170
Negative electrode, 133
Nucleation site, 85

Ohmic loss, 26
Over potential, 85, 86, 87, 88
Oxidation state, 130

Perfluorosurfactants, 87
Permeability, 5
Permittivity, 10
Plasticizer, 25
Polyethene, 88
Polymer radicals, 24
Porous-Type Membranes, 144
Positive Electrode, 136
Positive potential, 129

Pourbaix diagram, 84
Power, 177-181, 187, 191, 192,
Power plants, 238
Primary battery, 83
Proton migration, 84
Proton selective membrane, 84
Proton transport channels, 33
Proton-conducting channels, 28
Pumped hydro storage, 205

Quaternizedpolysulfone composite membranes (QNPSU), 88

Rechargeable batteries, 3
Redox, 51-59, 62-75
Redox active molecules, 4
Redox couple, 129
Redox Flow Batteries, 93, 162
Redox flow battery, 177-179, 182-184,186-188, 190, 193, 194,
Redox Flow Battery (RFB), 2
Redox reaction, 170
Reduction reaction, 132
Renewable energy, 178
Retention capability, 31
Reversible electrode, 84
Reversible energy storage, 83

Sealed lead-acid battery, 223
Secondary battery, 83
Self-discharge property, 11
Separator technology, 144
Silica gel, 88
Sodium-sulfur battery, 146
Solar, geothermal, wind energy, 83
Solvents, 11
Stability, 182, 185, 189
Stack efficiency, 130
State Laboratory PNNL, 129
State of charge, 140
Storage, 51-53, 56-58, 66, 73,
Super magnetic energy storage, 218
Supercapacitor, 215
Surface diffusion, 85
Surface Porosity, 132
Surfactants, 85
Swelling ratio, 10
SWNT, 83, 86, 87, 89

Tetra-n-butyl ammonium (TBA), 88
Tetrabutylammonium chloride (TBACl), 86
Tip current density, 85
Titanium dioxide, 136

Vanadium, 128
Vanadium batteries, 91
Vanadium pentaoxide, 137
Vanadium redox flow batteries, 102
Vanadium redox flow technology, 128
Voltage efficiency, 13

Yucel's method, 136

Zero gaps, 148
Zinc hydroxide, 84
Zinc rechargeable flow batteries, 93
Zinc sulphate, 85
Zinc-Nickel (Zn/Ni) Batteries, 97
Zinc-Sodium Hybrid Ion Batteries, 97
Zinc-Sulphur (Zn–S) Hybrid Battery, 97
Zn-Ce redox flow battery, 107
Zn–Br flow battery, 92, 95

Also of Interest

Check out these other forthcoming and published titles from Scrivener Publishing

Books by the same editor from Wiley-Scrivener

HANDBOOK OF WATER POLLUTION, Edited by Inamuddin, Tariq Altalhi, and Arwa Alrooqi, ISBN: 9781119904809. This handbook covers a wide range of water pollution issues and remedies. Researchers, students, and other governmental and non-governmental groups will all benefit from this informative guide. In detail, this book will include chapters in the areas of water pollution, analysis, and solutions.

ESSENTIAL OILS: Extraction Methods and Applications, Edited by Inamuddin, Tariq Altalhi, and Jorddy Neves Cruz, ISBN: 9781119829355. Written and edited by some of the world's top experts in the field, this exciting new volume provides up-to-date information on essential oils' chemical structure, therapeutic, and biological actions, and defines their functional uses and evaluates the advantages and disadvantages of their use in a variety of sectors.

PROTON EXCHANGE MEMBRANE FUEL CELLS: Electrochemical Methods and Computational Fluid Dynamics, Edited by Inamuddin, Omid Moradi, and Mohd Imran Ahamed, ISBN: 9781119829331. Edited by one of the most well-respected and prolific engineers in the world and his team, this book provides a comprehensive overview of hydrogen production, conversion, and storage, offering the scientific literature a comprehensive coverage of this important fuel.

MATERIALS FOR HYDROGEN PRODUCTION, CONVERSION, AND STORAGE, edited by Inamuddin, Tariq Altalhi, Sayed Mohammed Adnan, and Mohammed A. Amin, ISBN: 9781119829348. Edited by one of the most well-respected and prolific engineers in the world and his team, this book provides a comprehensive overview of hydrogen production, conversion, and storage, offering the scientific literature a comprehensive coverage of this important fuel.

FUNDAMENTALS OF SOLAR CELL DESIGN, edited by Inamuddin, Mohd Imran Ahamed, Rajender Boddula, and Mashallah Rezakazemi, ISBN: 9781119724704. Edited by one of the most well-respected and prolific engineers in the world and his team, this book provides a comprehensive overview of solar cells and explores the history of evolution and present scenarios of solar cell design, classification, properties, various semiconductor materials, thin films, wafer-scale, transparent solar cells, and other fundamentals of solar cell design.

Biofuel Cells, Edited by Inamuddin, Mohd Imran Ahamed, Rajender Boddula, and Mashallah Rezakazemi, ISBN 9781119724698. This book covers the most recent developments and offers a detailed overview of fundamentals, principles, mechanisms, properties, optimizing parameters, analytical characterization tools, various types of biofuel cells, edited by one of the most well-respected and prolific engineers in the world and his team.

Biodiesel Technology and Applications, Edited by Inamuddin, Mohd Imran Ahamed, Rajender Boddula, and Mashallah Rezakazemi, ISBN 9781119724643. This outstanding new volume provides a comprehensive overview on biodiesel technologies, covering a broad range of topics and practical applications, edited by one of the most well-respected and prolific engineers in the world and his team.

Applied Water Science Volume 1: Fundamentals and Applications, Edited by Inamuddin, Mohd Imran Ahamed, Rajender Boddula and Tauseef Ahmad Rangreez, ISBN 9781119724766. Edited by one of the most well-respected and prolific engineers in the world and his team, this is the first volume in a two-volume set that is the most thorough, up-to-date, and comprehensive volume on applied water science available today.

Applied Water Science Volume 2: Remediation Technologies, Edited by Inamuddin, Mohd Imran Ahamed, Rajender Boddula and Tauseef Ahmad Rangreez, ISBN 9781119724735. The second volume in a new two-volume set on applied water science, this book provides understanding, occurrence, identification, toxic effects and control of water pollutants in aquatic environment using green chemistry protocols.

Potassium-Ion Batteries: Materials and Applications, Edited by Inamuddin, Rajender Boddula, and Abdullah M. Asiri, ISBN 9781119661399. Edited by one of the most well-respected and prolific engineers in the world and his team, this is the most thorough, up-to-date, and comprehensive volume on potassium-ion batteries available today.

Rechargeable Batteries: History, Progress, and Applications, edited by Rajender Boddula, Inamuddin, Ramyakrishna Pothu, and Abdullah M. Asiri, ISBN 9781119661191. Edited by one of the most well-respected and prolific engineers in the world and his team, this is the most thorough, up-to-date, and comprehensive volume on rechargeable batteries available today.

Printed and bound by CPI Group (UK) Ltd, Croydon, CR0 4YY

17/09/2026

14964275-0001